Managing Wildlife in a Changing World

Edited by Jafari R. Kideghesho

Published in London, United Kingdom

IntechOpen

Supporting open minds since 2005

Managing Wildlife in a Changing World
http://dx.doi.org/10.5772/intechopen.81141
Edited by Jafari R. Kideghesho

Contributors
Marco Ferretti, Marco Zaccaroni, Vito Mazzarone, Luca Mattioli, Andrea Lenuzza, Paolo Banti, Massimo Taddei, Rocco Lopresti, Ludwig Krämer, Yakubu Joel Atuman, Caleb Kudi, P.A. Abdu, Aline Angelina Acosta, Lizaan De Necker, Wynand Malherbe, Marliese Truter, Louis Du Preez, Edward C. C. Netherlands, Ruan Gerber, Francois Retief, Reece Cronje Albert, Victor Wepener, Nico J. Smit, Dorothy Kagendo Kagendo, Eric Muchiri, Peter Gitonga, Esther Muthoni, Siew Shean Choong, Mimi Armiladiana Mohamad, Li Peng Tan, Ruhil Hayati Hamdan, Jafari R. Kideghesho, Jafari Kideghesho, O.O. Okubanjo, A. Abubakar, Rehema Abeli Shoo, Elizabeth Kamili Mtui, Julius Modest Kimaro, Neema Robert Kinabo, Gladys Joseph Lendii

First published in London, United Kingdom, 2021 by IntechOpen
IntechOpen is the global imprint of INTECHOPEN LIMITED, registered in England and Wales, registration number: 11086078, 5 Princes Gate Court, London, SW7 2QJ, United Kingdom
Printed in Croatia

British Library Cataloguing-in-Publication Data
A catalogue record for this book is available from the British Library

Additional hard and PDF copies can be obtained from orders@intechopen.com

Managing Wildlife in a Changing World
Edited by Jafari R. Kideghesho
p. cm.
Print ISBN 978-1-83880-975-1
Online ISBN 978-1-83880-976-8
eBook (PDF) ISBN 978-1-83880-977-5

Meet the editor

Professor Jafari R Kideghesho was born in Ugweno, Kilimanjaro in Northern Tanzania. He obtained his BSc in Agriculture from Sokoine University of Agriculture (SUA), Tanzania, in 1993; an MSc in Conservation Biology from Kent University, UK, in 1996; and a Ph.D. from Norwegian University of Science and Technology, in 2006. He started his career in wildlife management at the College of African Wildlife Management, Mweka, where he taught for six years before joining SUA in 1999. He served as a deputy director of the Wildlife Division in Tanzania's Ministry of Natural Resources and Tourism for two years (2012–2014). Dr. Kideghesho has been an active supporter of academic efforts within and outside Tanzania through teaching and serving as an external examiner at different universities. He has published more than fifty scientific articles in reputable journals and is an author and editor of numerous books. Currently, he is the rector at the College of African Wildlife Management, Mweka, Kilimanjaro.

Contents

Preface

The emerging and increasing socioeconomic, political, ecological, environmental, and technological changes occurring globally present a critical challenge to scientists, wildlife managers, and policy makers. Wildlife habitats are being degraded and fragmented as human demands for space for food production, infrastructure development, and settlements grow. Species are being pushed to the verge of extinction as their habitats are being degraded and humans attempt to meet their growing household and commercial needs from wildlife products. Wildlife species are further subjected to pre-emptive and retaliatory killing when they inflict economic losses and human mortalities. Climate change, diseases, and proliferation of invasive alien species are reducing the quality of habitats and affecting the population of wildlife species. Political unrest, civil wars, and terrorist acts in some regions have always disrupted the management operations of protected areas and paved a way for wildlife crimes. Reversing the declining and extinction trends that the world is experiencing today because of these factors calls for effective planning, innovations, and adoption of approaches that are developed through scientific realities. This book, *Managing Wildlife in a Changing World*, is comprised of eight chapters presenting issues and possible options for effective management of wildlife in a world where the changes are no longer speculative but rather are real and inevitable.

The introductory chapter by Jafari R. Kideghesho is a review of current status, trends, and drivers of various threats facing wildlife in the world. The author depicts an alarming rate of species loss and examines how different factors contribute to this loss. He underscores a need for effective planning, innovations, and adoption of approaches that are developed through scientific realities. He also proposes that deliberate efforts be taken to address the underlying and proximate causes responsible for the declining trends and extinction of wildlife species.

Chapter 2, "Conserving Freshwater Biodiversity in an African Sub-tropical Wetland: South Africa's Lower Phongolo River and Floodplain" by Acosta, et al., sheds light on the conservation and management issues of freshwater biodiversity in a highly diverse subtropical ecosystem. Based on a decade of survey conducted from 2010 to 2020 in the Phongolo River and Floodplain, this chapter highlights the current diversity of aquatic organisms (invertebrates, fishes, frogs, and their parasitic fauna), followed by an overview of their biological and physical stressors. Also addressed in this chapter are the current challenges in managing the aquatic biodiversity of this region and a way forward to conservation strategies.

In Chapter 3, "The Predicament of Macaque Conservation in Malaysia" Choong et al. provide some highlights on conservation of macaques species in Malaysia, pointing out that the proximity of macaques with human dwellings raises public health concern through the transmission of zoonotic diseases. The vulnerability of macaque species is increasing due to habitat loss, degradation, and fragmentation caused by forest clearing for plantation agriculture, selective logging, and increased network of roads because of urbanization. The existing false impression that all macaques are on equal ground and abundance in numbers has subjected these species to retaliatory and pre-emptive killing. The authors recommend increased scientific studies to understand the needs of these animals for continued survival and co-existence with humans and other animals in the ecosystem. Urgent efforts are required to preserve the natural habitats of the species along with creating public awareness on predicament of these species.

Chapter 4, "Diseases as Impediments to Livestock Production and Wildlife Conservation Goals" by Atuman et al., recognizes diseases transmission and pathogen spill-over at the wildlife–livestock interface as a growing concern to public and the scientific community due to impacts they exert on wildlife, livestock, and human health. The chapter describes the epidemiology of some viral infections (foot and mouth disease and rabies), bacterial infections (tuberculosis and brucellosis), and parasites (hemo and endoparasites) at the wildlife–livestock interface and potential impacts to livestock production and conservation goals. The authors recommend adoption of preventive measures that are geared towards improved disease surveillance among domestic and wild animals at the edges of protected areas using improved diagnostic techniques, vector control, and implementation of restrictions on anthropogenic animal movement, concomitant with public enlightenment campaign and behavioral change.

Chapter 5, "Interlinks between Wildlife and Domestic Cycles of *Echinococcus* spp. in Kenya" by Kagendo et al., presents cystic echinococcosis, a zoonotic disease of humans and animals, as a serious public health and economic problem in Sub-Saharan Africa, especially for pastoralists and nomadic communities. The authors attribute failure to control the disease to lack of resources and limited knowledge about epidemiology. They advocate for more research on problems related to wildlife diseases, determining the presence of such diseases, their prevalence, and their influence on wildlife conservation to improve disease outbreak preparedness.

Chapter 6, "Wildlife Management Areas in Tanzania: Vulnerability and Survival Amidst COVID-19" by Shoo et al., uses the five Wildlife Management Areas (WMAs) located in Northern Tanzania to show the negative impacts that WMAs have suffered following the outbreak of COVID-19 and, consequently, disruption of flow of revenues from tourism. WMAs, established as intervention to safeguard wildlife and their habitats outside the core protected areas, provide an opportunity for local communities to manage and benefit from these areas and therefore their poor performance has watered down conservation efforts and local livelihoods. To avoid similar impacts in the future, the authors recommend the creation of local mechanisms for revenue acquisition that are more resilient to global shocks, diversifying revenue-generating options within WMAs, and putting in place the right funding model that would warrant WMAs' sustainability.

In Chapter 7, "A Step Change in Wild Boar Management in Tuscany Region, Central Italy," Zaccaroni et al. describe the challenges associated with managing wild ungulates. Increased population of the species has intensified human-wildlife conflicts and economic burden to the extent that the traditional management approach can hardly cope with the challenge. The authors outline some strategies to complement the existing traditional management approach.

In Chapter 8, "Managing Invasive Alien Species by the European Union: Lessons Learnt," by Ludwig Kramer sheds light on the fight against invasive alien species within the European Union. In 2014, the EU adopted a regulation to identify and manage invasive alien species. This chapter discusses the regulation and its monitoring, highlighting the lessons learned from the cooperation of the different states in the EU.

This book is in no way exhaustive in terms of presenting all emerging issues and trends with implications on wildlife conservation. Only four issues, among others, are subjects of this book: habitat management, invasive species, human–wildlife conflicts, and diseases. Other issues including poaching and illegal wildlife trade, climate change, poverty, corruption, political unrest, urbanization, and emerging of economic opportunities and development projects have not been addressed. This is not because they are less damaging to wildlife, but because the scope of this book was limited to four issues only. However, the book alerts managers, policy makers, and the international community on the imminent danger facing wildlife resources and a need to pay more attention on these issues and trends. The book also inspires academicians and researchers to play their roles in providing adequate scientific data and recommendations in view of improving evidence-based decisions for practical solutions that can abate and reverse the current declining trends on wildlife and biodiversity resources.

Jafari R. Kideghesho
College of African Wildlife Management, Mweka,
Moshi, Tanzania

Chapter 1

Introductory Chapter: Managing Wildlife in a Changing World - Trends, Drivers and the Way Forward

Jafari R. Kideghesho

1. Introduction

The wildlife managers, scientists and policy makers globally, are striving to ensure the survival of wildlife resources in the face of rapid changes in socio-economic, ecological, political and technological aspects. One of the key and popular strategies that have been adopted to conserve wildlife species is the establishment of different categories of protected areas. A protected area is defined as "geographical space, recognized, dedicated and managed, through legal or other effective means, to achieve the long term *conservation* of nature with associated ecosystem services and cultural values" [1]. Since the establishment of Yellowstone National Park in 1872 - the first protected area in the world - a number of protected areas has been growing over time. In 2010, there were about 161,000 protected areas [2]. *The 2016 Protected Planet Report* [3] indicated an increase to 202,470 protected areas spanning about 20 million square kilometers (15% of the world's land, excluding Antarctica). As of August 2020, the number increased beyond 260,000 [4].

Establishment of protected areas has been complemented with many other strategies. Local and global commitment have been apparent through enacting and enforcement of numerous laws; provision of alternative livelihood strategies as a substitution to ecologically destructive activities; and supporting community-based conservation programmes including conservation education and awareness creation, benefit-sharing schemes, among others.

Local, regional and international instruments have been established to spearhead the conservation efforts and stem the causes of species loss. The popular international instruments, among others, include Convention on International Trade in Endangered Species of Flora and Fauna – CITES (year of entry into force: 1975), Convention of Biological Diversity -CBD (1993), Convention on Wetlands -popularly known as the Ramsar Convention (1971), Convention on the Conservation of migratory species of wild animals -CMS or Bonn Convention (1975) and World Heritage Convention - WHC (1972). The regional instruments include The Lusaka Agreement on Co-operative Enforcement Operations Directed at Illegal Trade in Wild Fauna and Flora; The Southern Africa Development Community Protocol on Wildlife Conservation and Law Enforcement; The Convention on Conservation of Nature in the South Pacific; ASEAN Agreement on the Conservation of Nature and Natural Resources' and Convention on the Conservation of European Wildlife and Natural Habitats (Bern Convention).

2. Declining trends and drivers

Despite efforts which are being devoted locally, regionally and globally to curb the threats facing wildlife, it is increasingly becoming evident that these efforts are not matching the rates of the threats. The UN Report titled *Nature's Dangerous Decline 'Unprecedented'; Species Extinction Rates 'Accelerating'* released by the Intergovernmental Science-Policy Platform on Biodiversity and Ecosystem Services (IPBES), indicates that roughly one million animal and plant species are currently threatened with extinction [5]. According to the Report, most major land-based habitats have lost at least 20% of the average abundance of their native species since 1900. Over 40% of amphibian species and 33% all marine mammals, respectively, are threatened with extinction. Since the 16th century about 680 vertebrate species have been driven to extinction [5]. Furthermore, the recent report of IUCN *Redlist of Threatened Species* has revealed that, out of 158,908 species of vertebrates assessed, 35,300 (equal to 28%) are threatened with extinction [6]. Among other species, the declining trends involve the umbrella, keystone and charismatic species such as black rhino (*Bicornis diceros*), African elephant (*Loxodonta africana*), tiger (*Panthera tigris*), African lion (*Panthera leo*) and leopard (*Panthera pardus*) [7–12].

The declining trends of wildlife species and biodiversity in general, is a function of multiple factors (**Figure 1**) associated with socio-economic, ecological, technological or political changes. Growing human demand for food, timber, water and space is increasingly causing habitat loss and deterioration and, thus, subjecting many species to a risk of extinction [12]. According to IPBES Report (5), about 47% of natural ecosystems have declined globally and over 9% of the world's estimated 5.9 million terrestrial species have insufficient habitat for long term survival without efforts to restore the degraded habitats. Reduced home ranges for different wildlife species and blockage of corridors are exacerbating property damage and human mortalities and, therefore, inciting pre-emptive or retaliatory killing [12, 13]. Likewise, poaching and illegal wildlife trade has accelerated declining rates to species of high economic value such as rhino, elephant, pangolins and tiger [14–19].

Human population growth is the main driving force behind most of the threats facing wildlife species. It is linked to current trends of invasive species, climate change, wildlife crime, pollution, and habitat loss and human-wildlife conflicts. The current world population of 7.8 billion [20] is projected to reach 8.6 billion in 2030;

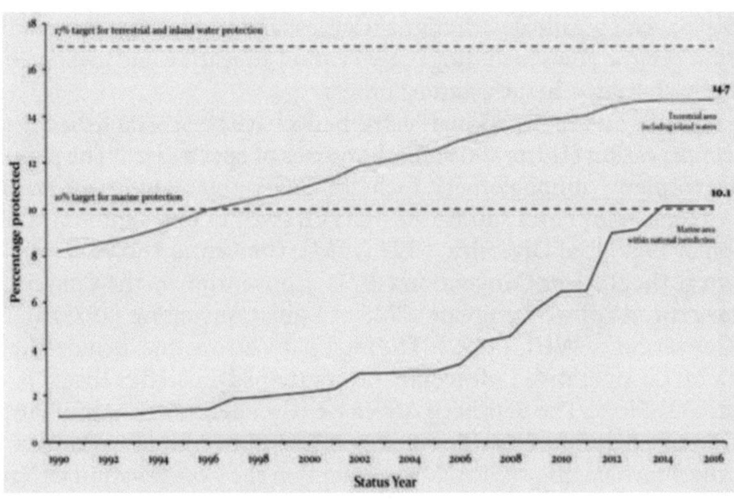

Figure 1.
Increasing trends of protected areas globally from 1990 to 2016 [3].

9.8 billion in 2050 and 11.2 billion in 2100 [21]. These projections signify that demand and consumption for resources such as food, fuel, timber and space will increase significantly at the expense of wildlife species as more land will be transformed into human settlements and infrastructures. It is estimated that humans and domesticated livestock account for about 36 and 60 percent of the biomass of all mammals on Earth, respectively, while wild mammals have declined to only 4 percent [22].

Blockage of wildlife corridors and, subsequently, habitats fragmentation is rendering many protected areas isolated as ecological islands [23–25]. Disruption of the ecological linkage between different animal populations, consequently, reduces the genetic variability due to inbreeding depression [12–27]. Furthermore, loss of wildlife corridors and dispersal areas compromise their roles in minimizing human-wildlife conflicts, provision of alternative foraging or breeding grounds and serving as a refuge against adverse weather conditions [27].

Technological advancement, improved accessibility to remote areas and availability of markets for wildlife products have also worsened the destruction of habitats and depletion of wildlife species [28–31]. Emerging of new economic opportunities and increasing need for development are giving rise to adoption of policy choices that are economically rewarding but ecologically damaging such as mining, industrial agriculture and construction of infrastructures [32–37]. Political unrest and associated effects such as insecurity, proliferation of firearms, influx of refugees and disruption of operations of protected areas has contributed enormously to the increased decimation of wildlife populations and destruction of habitats [38–40].

Climate change is increasingly featuring as one of the important global agenda impacting nature and human life. The main driver of climate change is destruction of nature through human actions such as increased use of fossil fuels, deforestation and intensive agriculture. Food insecurity and income poverty ensuing as a result of climate change leave people with limited livelihood options, a situation that may prompt engagement of people in poaching and habitat destruction [41]. Climate change accelerates habitat threshold and increases risk to species in fragmented habitats [42]. Climate change is confirmed to facilitate the increase of wildlife diseases [43–45], spread of invasive species [45–47] and escalation of human-wildlife conflicts [12, 48].

In recent decades the number of invasive alien species has kept on growing and thus posing a serious threat to native wildlife species [49–52]. Nearly 40% of introductions of invasive alien species in the past two centuries occurred between 1970 and 2014 [46]. According to Seebens et al. [53] the projected overall increase of alien species between 2005 and 2050 was 36% per continent. The challenge is increasing in tandem with the changes which are taking place globally such as international trade; global transport of goods, population changes, migration, pollution, tourism, recreation, climate change and economic development such as land use and energy consumption [54–59]. Invasive alien species present threat to native species by out-competing them for food and other resources, destroying their habitats, introducing diseases, thwarting reproduction of native species, preying on native species or killing the young of native species [60–63]. The proportion of threatened or endangered species facing a risk of extinction because of invasive species is estimated at 42% [64]. Among other threats, Invasive species are currently threatening 27% of the globally threatened terrestrial species of mammals, birds and reptiles included on the IUCN Red List, and 40% of the critically endangered species, in particular [52].

Human population growth, urbanization, habitat loss, poverty, climate change and improved conservation measures, among other drivers, are intensifying human-wildlife conflicts in many parts of the world. Numerous wildlife species are subjected to risk as human-wildlife conflicts intensify. Economic loss and mortalities caused by problem and dangerous animals incite retaliation in form of killing and destroying wildlife and their habitats [48]. Similarly, population growth, destruction of ecosystems, climate change along with increased human activities such as industrialization, mining, water-waste, metal refining and the burning of fossil fuels are exacerbating pollution in form of synthetic chemicals, oil spills, toxic metals and acid rain. The documented impacts of pollutants to wildlife species include immediate deaths, habitat destruction, reduced or impaired reproduction, cancer, neurological damage, liver damage, muscle atrophy and immune suppression to diseases [65–67].

Increasing human impacts on ecosystems, climate change, invasive alien species and pollution are attributed to emerging and re-emerging of diseases affecting humans and non-human species [65, 68–71]. Human-induced changes on land use and land cover through modification of natural habitats are responsible for over 50% of the emerging zoonoses [72–76]. Given the global human population growth and deforestation rate, estimated at 10 million hectares per annum [76], it is indisputably that the risk of animal-to-human diseases transmission will increase talong with increased proximity of humans and livestock to natural habitats. Diseases have both direct and indirect impacts on wildlife species. Direct impacts involve effect of a disease on the health of animal species which can subsequently lead to deaths. Examples of diseases with direct impact on the health of animals are anthrax and Canine Distemper Virus Disease. Indirect impacts are impacts which disrupt the management interventions of the species and habitats. For instance, emerging of pandemics (e.g., Ebola and COVID-19) and, consequently, imposition of travel restrictions and lockdown had reduced revenues from tourism and impacted the livelihood of many people. This has denied conservation authorities adequate resources for conservation and, therefore, subjecting wildlife species to risks including poaching [75].

3. The way forward

Reversing the declining and extinction trends that the world is experiencing today calls for effective planning, innovations and adoption of approaches that are developed through scientific realities. Understanding of the factors and the mechanisms in which they influence the survival of wildlife is critical in devising the mitigation against the current challenges facing wildlife. Deliberate efforts are required to address the underlying and proximate causes of the declining trends and extinction of wildlife species. This book "*Managing Wildlife in a Changing World*" presents issues and possible options for effective management of wildlife in a world where the changes are no longer speculative, but a reality and inevitable. By reading the Book, you will realize that not all conservation issues require biological solutions. Sociological approaches are essential as most of the current challenges are anthropogenic in nature. Issues such as poverty, human population growth, human-wildlife conflicts, illegal use of resources, habitat loss, proliferation of invasive species and diseases, among others, call for informed policies, public awareness, wide stakeholder involvement in planning, decision-making and implementation of conservation measures.

Author details

Jafari R. Kideghesho
College of African Wildlife Management, Moshi, Tanzania

*Address all correspondence to: kideghesho@yahoo.com

References

[1] Borrini-Feyerabend, G., N. Dudley, T. Jaeger, B. Lassen, N. Pathak Broome, A. Phillips and T. Sandwith (2013). *Governance of Protected Area: From understanding to action*. IUCN. ISBN 978-2-8317-1608-4.

[2] Soutullo, A. (2010). Extent of the global network of terrestrial protected areas. *Wildlife Biology*, 24 (2): 362-363

[3] UNEP-WCMC and IUCN (2016). *Protected Planet Report 2016.* UNEP-WCMC and IUCN: Cambridge UK and Gland, Switzerland.

[4] *UNEP-WCMC (2020). August 2020 update of the WDPA". Protected Planet. UNEP-*WCMC. https://www.protectedplanet.net/c/monthly-updates/2020/august-2020-update-of-the-wdpa. Retrieved 19 August 2020.

[5] IPBES (2019): *Summary for policymakers of the global assessment report on biodiversity and ecosystem services of the Intergovernmental Science-Policy Platform on Biodiversity and Ecosystem Services.* IPBES secretariat, Bonn, Germany. 56 pages.

[6] IUCN 2021. *The IUCN Red List of Threatened Species*. Version 2020-3. https://www.iucnredlist.org>

[7] Riggio J., Jacobson A, Dollar L., Bauer H., Becker M., Dickman A., et. al. (2012). The size of savannah Africa: a lion's (*Panthera leo*) view. *Biodiversity Conservation,* DOI: 10.1007/s10531-012-0381-4.

[8] World Bank (2005). Going, Going, Gone: *The Illegal Trade in Wildlife in East and Southeast Asia*. Environment and Social Development East Asia and Pacific Region Discussion Paper. www.worldbank.org/biodiversity & www.worldbank.org/eapenvironment.

[9] WWF (2018). Second-biggest direct threat to species after habitat destruction. https://wwf.panda.org/discover/our_focus/wildlife_practice/problems/illegal_trade/

[10] Jacobson AP, Gerngross P, Lemeris Jr. JR, Schoonover RF, Anco C, Breitenmoser-Würsten et al. (2016). Leopard (*Panthera pardus*) status, distribution, and the research efforts across its range. *Peer Journal* 4:e1974 https://doi.org/10.7717/peerj.1974.

[11] Stephens P.A. (2015).Land sparing, land sharing, and the fate of Africa's lions. *PNAS* 112 (48):14753-14754

[12] Kideghesho JR, Mseja GA, Dulle HI and Nyakunga OC. (2020). Conservation of large mammals in the face of increasing human population and urbanization in Tanzania. Can Artificial Nest Boxes be Effective Conservation Tools? In Martin EH, Hadfield LA, Melubo K, Jensen RR, Durrant J0, Hardin PJ & Weisler L. (Eds.). *Protected Areas in Northern Tanzania,* pp.157-179. Springer Nature Switzerland AG 2020

[13] Trinkel, M. and Angelici F.M. (2016). The Decline in the Lion Population in Africa and Possible Mitigation Measures. In F.M. Angelici (Ed.), *Problematic Wildlife*, DOI 10.1007/978-3-319-22246-2_

[14] Moneron S, Armstrong A and Newton D. (2020). *The People Beyond the Poaching: Interviews With Convicted Offenders in South Africa.* TRAFFIC International, Cambridge, United Kingdom. https://www.traffic.org/publications/reports/the-people-beyond-the-poaching/

[15] WWF (2018). The status of African elephants (*Loxodonta Africana*). https://www.worldwildlife.org/magazine/issues/winter-2018/articles/the-status-of-african-elephants

[16] EIA (2021). *Illegal trade Seizures: Pangolins*. Environmental Investigation Agency UK

[17] Nuwer, R. (2020). Illegal trade in pangolins keeps growing as criminal networks expand. *National Geographic Magazine,* https://www.nationalgeographic.com/animals/article/pangolin-scale-trade-shipments-growing

[18] Cooke J. G. and Schilckeisen D. (2015). *Africa's Wildlife Poaching and Trafficking Crisis. The Need for Urgent Action*. CSIS (Center for Strategic and International Studies). https://www.csis.org/analysis/africas-wildlife-poaching-and-trafficking-crisis

[19] Conciatore, J. (2019). Up to 2.7 million pangolins are poached every year for scales and meat. *African Wildlife Foundation*, https://www.awf.org/blog/27-million-pangolins-are-poached-every-year-scales-and-meat

[20] PRB (2020). *2020 World Population Data Sheet Shows Older Populations Growing, Total Fertility Rates Declining.* https://www.prb.org/2020-world-population-data-sheet/

[21] United Nations Department of Economic and Social Affairs Population Dynamics (2019). *World Population Prospects 2019.* https://population.un.org/wpp/

[22] The biomass distribution on Earth. (2018). *Proceedings of the National Academy of Sciences of the United States of America*, 115 (25):6506-6511.

[23] Quinn, J. F. and S. P. Harrison. 1988. Effects of habitat fragmentation and isolation on species richness: evidence from biogeographic patterns. *Oecologia* 75: 132-140.

[24] Robinson G.R. and Quinn J.F. (1992) Habitat fragmentation, species diversity, extinction, and design of nature reserves. In: Jain S.K., Botsford L.W. (eds) *Applied Population Biology. Monographiae Biologicae*, 67. Springer, Dordrecht. https://doi.org/10.1007/978-0-585-32911-6_10

[25] Kideghesho, J. R.; Rija, A. A.; Mwamende, K. A. and Selemani, I. (2013). Emerging issues and challenges in conservation of biodiversity in the rangelands of Tanzania. *Nature Conservation* 6: 1-29. doi: 10.3897/natureconservation.6.5407

[26] Kideghesho, J.R. (2015) Realities on deforestation in Tanzania: trends, drivers, implications and the way forward. In Zlatic M (Ed) *Precious Forests - Precious Earth*. ISBN 978-953-51-2175-6, INTECH Open Science/Open minds

[27] Kideghesho J.R., Nyahongo, J.W., Hassan, S.N., Tarimo, T.C. and Mbije, N.E. (2006). Factors and ecological impacts of wildlife habitat destruction in the Serengeti Ecosystem in Northern Tanzania. *African Journal of Environmental Assessment and Management,* 11:17-32 http://www.ajeam-ragee.org/

[28] Kioko, J., Kiffner, C., Jenkins, N. and Collinson, W.J. 2014. Wildlife road kill patterns on a major highway in northern Tanzania. *African Zoology* 50: 17-22

[29] Coffin, A. W. (2007). From road kill to road ecology: A review of the ecological effects of roads. *Journal of Transport Geography,* 15:396-406.

[30] Maschio, G. F., Santos-Costa, M. C., Prudente, A. L. (2016). Road-kills of snakes in a tropical rainforest in the central Amazon basin, Brazil. *South American Journal of Herpetology,* 11:46-53.

[31] Loss, S.R., Will, T. and Marra, P.P. (2018). Refining estimates of bird collision and electrocution mortality at power lines in the United States. PLoS One. 2014;9(7):e101565.

[32] Giam, X., Olden, J.D. & Simberloff, D. Impact of coal mining on stream biodiversity in the US and its regulatory implications. *Nature Sustainability* 1:176-183 https://doi.org/10.1038/s41893-018-0048-6

[33] G.W. Fernandes, F.F. Goulart, B.D. Ranieri, M.S. Coelho, K. Dales, N. Boesche, M. Bustamante, F.A. et al. (2016). Deep into the mud: ecological and socio-economic impacts of the dam breach in Mariana, Brazil. *Perspectives of Ecological Conservation (Natureza & Conservação),* 14:35-45

[34] BirdLife International (2008) Campaign to save Mabira Forest in Uganda from sugarcane plantation for biofuels. Downloaded from http://www.birdlife.org on 15/11/2020

[35] L.J. Sonter, D.J. Barrett, B.S. Soares-Filho. Offsetting the impacts of mining to achieve no net loss of native vegetation. *Wildlife Biology*, 28 (2014), pp. 1068-1076

[36] Redmond, I. 2001. *Coltan boom, gorilla bust: the impact of coltan mining on gorillas and other wildlife in eastern DR Congo*. Report for The Dian Fossey Fund International, Atlanta , and The Born Free Foundation, Horsham , Rhode Island

[37] Michael B. Mascia & Sharon Pailler (2011). Protected area downgrading, downsizing, and degazettement(PADDD) and its conservation implications. *Conservation Letters* 9-20

[38] Dudley, J. P., J. R. Ginsberg, A. J. Plumptre, J. A. Hart, and L. C. Campos. 2002. Effects of war and civil strife on wildlife and wildlife habitats. *Conservation Biology* 16: 319– 329.

[39] Jeffrey A. McNeely (2018) *Wildlife at War in Angola: The rise and fall of an African Eden, Transactions of the Royal Society of South Africa,* 73:2, 202-204, DOI: 10.1080/0035919X.2017.1409494

[40] Braga-Pereira F, Peres CA, Campos-Silva JV, Santos CV, Alves RRN (2020). Warfare-induced mammal population declines in Southwestern Africa are mediated by species life history, habitat type and hunter preferences. *Scientific Reports,* 10(1): 15428 https://doi.org/10.1038/s41598-020-71501-0

[41] Kideghesho and Msuya (2013). Managing the wildlife protected areas in the face of global economic recession, HIV/AIDS pandemic, political instability and climate change: Experience of Tanzania. In Sladonja, B. (Ed.). *Protected Areas Management*. INTECH Open Science/Open minds, pp 65-91. ISBN: 980-953-307-448-6

[42] Travis, J.M.J. (2003). Climate change and habitat destruction: A deadly anthropogenic cocktail. *Proceedings of Royal Society of London.* B.270467-473. http://doi.org/10.1098/rspb.2002.2246

[43] Harvell, D., Altizer, S., Cattadori, I.M., Harrington, L. and Weil, E. (2009). Climate change and wildlife diseases: When does the host matter the most? *Ecology* 90(4):912-920i

[44] Lips, K. R., F. Brem, R. Brenes, J. D. Reeve, R. A. Alford, J.Voyles, C. Carey, L. Livo, A. P. Pessier, and J. P. Collins.2006. Emerging infectious disease and the loss of biodiversityin a Neotropical amphibian community. *Proceedings of theNational Academy of Sciences* (USA) 103:3165-3170.

[45] Lips, K. R., J. Diffendorfer, J. R. Mendelson, and M. W. Sears. 2008. Riding the wave: reconciling the roles of disease and climate change in amphibian declines. *PLoS Biology* 6(3):e72.

[46] Seebens, H., Blackburn, T., Dyer, E. *et al.* [2017]. No saturation in the accumulation of alien species worldwide. *Nature Communication* 8, 14435. https://doi.org/10.1038/ncomms14435

[47] Hellmann, J.J., Byers, J.E., Bierwagen, B.G., and Dukes, J.S. (2008). Five potential consequences of climate change for invasive species. *Conservation Biology* 22(3):534-543

[48] Otiang'a-Owiti, G. E., Nyamasyo, S., EMalel, E., and Onyuro, R. (2011). Impact of climate change on human-wildlife conflicts in East Africa. *Kenya Veterinarian* 35:103-110. Available online at: https://www.ajol.info/index.php/kenvet/article/view/87548

[49] Atkinson, I.A.E. (1996). Introductions of wildlife as a cause of species extinctions. *Wildlife Biology*, 2(3):135-141 (1996). https://doi.org/10.2981/wlb.1996.011

[50] Linders, T.E.W., Schaffner, U., Eschen, R., Abebe, A., Choge, S.K., Nigatu, L, Mbaabu, P.R., Shiferaw, H, and Allan, E. (2019). Direct and indirect effects of invasive species: Biodiversity loss is a major mechanism by which an invasive tree affects ecosystem functioning. *Journal of Ecology,* 107(6): 2660-2672

[51] Bellard, C., Genovesi, P., Jeschke, J.M. (2016). Global patterns in threats to vertebrates by biological invasions. *Proceedings of the Royal Society B: Biological Sciences* 283 (1823) (2016), p. 20152454

[52] Bellard, C., Cassey, P. and Blackburn, T.M. (2016). Alien species as a driver of recent extinctions. *Biological Letters* 12 (2016), p. 20150623

[53] Seebens, H., Bacher, S., Blackburn, T.M., Capinha, C., Dawson, W., Dullinger, S., Genovesi, P., Hulme, P.E., Kleunen, M., Kühn, I., Jeschke, J.M., Lenzner, B., Liebhold, A.M., Pattison, Z., Pergl, J., Pyšek, P., Winter, M. and Essl, F. (2020). Projecting the continental accumulation of alien species through to 2050. *Global Change Biology* 27(1):16. DOI: 10.1111/gcb.15333

[54] Mueller, J. M. & Hellmann, J. J. (2008).An assessment of invasion risk from assisted migration. *Conservation Biology* 22:562-567

[55] Hulme, P. E. Trade, transport and trouble: managing invasive species pathways in an era of globalization. *Journal of Applied Ecology,* 46:10-18 (2009).

[56] Banks, N. C., Paini, D. R., Bayliss, K. L. & Hodda, M. (2015).The role of global trade and transport network topology in the human-mediated dispersal of alien species. *Ecological Letters* 18:188-199

[57] Skarpaas, O. & Økland, B. Timber. (2009). import and the risk of forest pest introductions. *Journal of Applied Ecology* 46, 55-63

[58] McNeill, M. et al. (2011). Transportation of non-indigenous species via soil on international aircraft passengers' footwear. *Biological Invasions* 13:2799-2815

[59] Huang, Z., Das, A., Qiu, Y. & Tatem, A. (2012).Web-based GIS: the vector-borne disease airline importation risk (VBD-AIR) tool. *International Journal of Health Geogaphy* 11, 33

[60] Witte, F., Goldschmidt, T., Wanink, J. *et al.* (1992). The destruction of an endemic species flock: quantitative data on the decline of the haplochromine cichlids of Lake Victoria. *Environment Biology of Fish* 34, 1-28 https://doi.org/10.1007/BF00004782

[61] Savidge, J.A. (1987). Extinction of an Island Forest Avifauna by an Introduced Snake. *Ecology,* 68(3): 660-668

[62] Aloo, P. A., Njiru, J., Balirwa, J. S., and Nyamweya, C. S. 2017. Impacts of Nile Perch, *Lates niloticus*, introduction on the ecology, economy and conservation of Lake Victoria, East Africa. *Lakes and Reservoirs: Research*

and Management. 22(4): 320-333. doi: https://doi.org/10.1111/lre.12192

[63] Wiles, G.J., Bart, J., Beck, R.E. Jr., and Aguon, C.F. (2003). Impacts of the Brown Tree Snake: Patterns of Decline and Species Persistence in Guam's Avifauna. *Conservation Biology,* 17(5):1350-1360

[64] NWF. The National Wildlife Federation -Invasive Species, Available at: https://www.nwf.org/Educational-Resources/Wildlife-Guide/Threats-to-Wildlife/Invasive-Species

[65] Acevedo-Whitehouse, K. and Duffus, A.L.J. (2009) Effects of environmental change on wildlife health. *Philosophical Transactions: Biological Sciences,* 27; 364(1534): 3429-3438. doi: 10.1098/rstb.2009.0128

[66] Newman, J.R. (1979). Effects of industrial air pollution on wildlife. *Biological Conservation,* 15(3):181-190

[67] Preeti, Reen,J.K., Thakur,M., Suman, M. and Kumar, R. (2018). Consequences of pollution in wildlife: A review. *The Pharma Innovation Journal,* 7(4): 94-102

[68] Harvell, D., Altizer, S., Cattadori, I.M., Harrington, L. and Weil, E. (2009). Climate change and wildlife diseases: When does the host matter the most? *Ecology,* 90(4):912-920

[69] Lips, K. R., F. Brem, R. Brenes, J. D. Reeve, R. A. Alford, J.Voyles, C. Carey, L. Livo, A. P. Pessier, and J. P. Collins. 2006. Emerging infectious disease and the loss of biodiversity in a Neotropical amphibian community. *Proceedings of theNational Academy of Sciences* (USA) 103:3165-3170.

[70] Lips, K. R., J. Diffendorfer, J. R. Mendelson, and M. W. Sears. 2008. Riding the wave: reconciling the roles of disease and climate change in amphibian declines. *PLoS Biology* 6(3):e72.

[71] McMichael, A., and M. J. Bouma. 2000. Global change, invasive species and human health. Pages 191-210 in H.Mooney and R. Hobbs, (Eds). *The Impact of Global Change on Invasive Species.* Island Press, Covelo, California, USA.

[72] Loh, E.H., Zambrana-Torrelio, C., Olival, K.J., Bogich, T.L., Johnson, C.K., Mazet, J.A.K., Karesh, W. and Daszak, P. (2015). Targeting transmission pathways for emerging zoonotic disease surveillance and control. *Vector Borne and Zoonotic Diseases,* 15(7), 432-437.

[73] WWF (2020).*The loss of nature and rise of pandemics: protecting human and planetary health.*WWF International, Gland, Switzerland. 19pp

[74] Kaneda, T. and Greenbaum, C. (2020). How demographic changes make us more vulnerable to pandemics like the coronavirus. Available at: https://www.prb.org/how-demographic-changes-make-us-more-vulnerable-to-pandemics-like-the-coronavirus/

[75] Kideghesho, J.R., Kimaro, H.S., Kisingo, A.W. and Mayengo, G.M. (2021). Will Tanzania's wildlife sector survive the COVID-19 pandemic? *Tropical Conservation Science,* https://doi.org/10.1177/19400829211012682

[76] Food and Agriculture Organization of the United Nations (2020). *Global Forest Resources Assessment 2020*. Rome: FAO. https://doi.org/10.4060/ ca8753en

Chapter 2

Conserving Freshwater Biodiversity in an African Subtropical Wetland: South Africa's Lower Phongolo River and Floodplain

Aline Angelina Acosta, Edward C. Netherlands, Francois Retief, Lizaan de Necker, Louis du Preez, Marliese Truter, Reece Alberts, Ruan Gerber, Victor Wepener, Wynand Malherbe and Nico J. Smit

Abstract

Freshwater biodiversity is under constant threat from a range of anthropogenic stressors. Using South Africa's Phongolo River and floodplain (PRF) as a study case, the aim of this chapter is to provide an overview of the conservation and management of freshwater biodiversity in a highly diverse subtropical ecosystem. The PRF is the largest floodplain system in South Africa which is severely threatened by irregularly controlled flood releases from a large upstream dam, prolonged drought, deteriorating water quality, organic pollutants and the increasing dependence of the local communities. Based on a decade of survey of the PRF conducted from 2010 to 2020, this chapter highlights the current diversity of aquatic organisms (invertebrates, fishes, frogs and their parasitic fauna), followed by an overview of their biological and physical stressors. The current challenges in the management of the aquatic biodiversity of this region and a way forward to conservation strategies are also addressed in this chapter.

Keywords: conservation, aquatic organisms, biological stressors, physical stressors, management

1. Introduction

The Phongolo River (PR) originates in the South African Mpumalanga Province from where it flows first eastwards, before turning north through the Ubombo mountain ranges in South Africa's northern KwaZulu-Natal (KZN) [1]. The lower Phongolo River and associated floodplain (PRF) starts from where the river exits a gorge in the Ubombo mountains, which is known as the Pongolapoort, for approximately 80 km downstream to the confluence of the PR and Usuthu River (UR) at the South Africa/Mozambique border [2] (**Figure 1**). The PRF is about 10,000 increasing to 13,000 ha in full inundation and is characterised by its permanent and

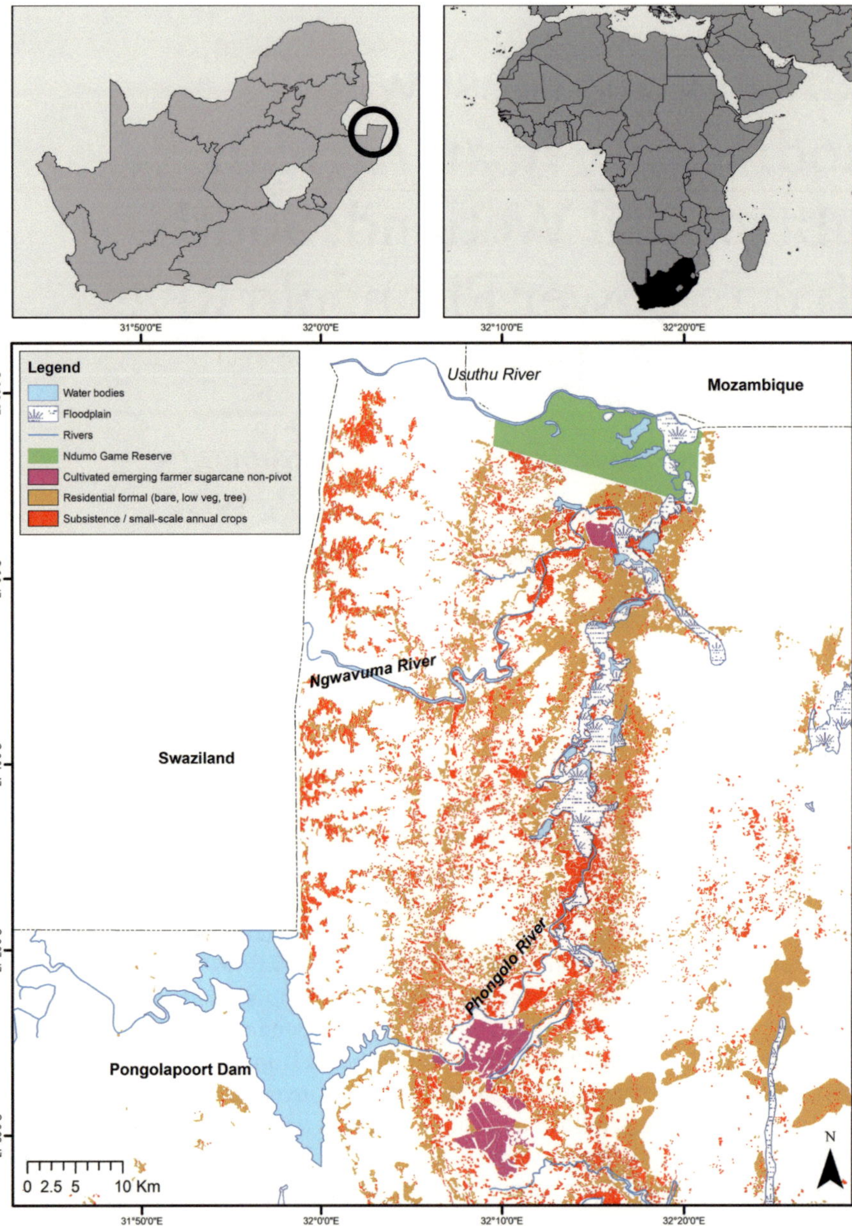

Figure 1.
Map of the Pongolapoort dam, the lower Phongolo River and its associated floodplain in North-Eastern South Africa.

temporary floodplain depressions (from here on further referred to as pans) that have either permanent, intermitted or no connectivity with the river. Natural flow regime determines the heterogeneity of the floodplain habitats. Pans that are river-fed function as spawning and breeding sites and as foraging areas for migratory species. The temporary pans are also known to harbour endemic species, therefore presenting a unique diversity. The connection between the river and floodplain also enables the exchange of organic matter, nutrients and aquatic vegetation. Thus, the flood pulse plays a key role in the health of the river ecosystem and productivity of the wetlands. The PRF constitutes one of the largest and the most biodiverse floodplain systems in South Africa.

The previously mentioned gorge (Pongolapoort) in the Ubombo mountains provided the perfect structural position for the building of a dam and therefore it was no surprise that the building of a dam in the PR (**Figure 2A, B**), with the main aim to supply irrigation water for sugarcane and cotton crops, was commissioned in the 1960s. The Pongolapoort Dam (also referred to as the Jozini Dam) was completed in 1973, and although research on its possible effect on the PRF started before the completion of the dam, there have been continuous concerns over the past almost 50 years about the real impact thereof on the PRF. In order to mitigate the dam's impacts on the ecological integrity of and ecosystem services provided by the floodplain, a controlled flood release regime was developed to simulate natural flooding [3]; however, this was never fully implemented [1, 4]. The fragile ecosystem of the PRF has also, over the past 40 years, been further impacted by extreme climatic events such as a cyclone (1984) and severe droughts (1981–1983 and 2016–2020) [1].

The only officially protected section of the PRF lays within the 102 km^2 Ndumo Game Reserve (NGR) that was proclaimed in 1924 (**Figure 1**). In 1997, the game reserve was also declared a Ramsar site under the Wetlands of International Importance convention [2]. In addition to the PR and UR sections that fall inside the reserve, the NGR also provides protection to five distinct wetland types ranging between fresh, brackish and saline and permanent and intermittent rivers, lakes, pools and riparian/gallery forests. One of the largest water bodies and also the only naturally saline lake (approximately 5000 $\mu S \cdot cm^{-1}$) within NGR is Lake Nyamithi (183.4 ha) (**Figure 2C**). Nyamithi receives water largely from the PR through the annually controlled flood release from the Pongolapoort Dam, through rainfall during the wet season (summer) from its own small catchment and natural floods from the UR. Although the reserve is almost 100 years old, very little has been accomplished in terms of research since the reserve's establishment, and it is unknown whether the water that NGR receives is of a quality and quantity to support and protect this biodiversity hotspot.

Figure 2.
(A) View of the dam wall from the Phongolo River; (B) downstream view from the dam wall; (C) Lake Nyamithi and (D) example of activity of people directly dependent on the floodplains for cultural and provisioning ecosystem services.

Figure 3.
View of the Phongolo River flowing North with Ndumo Game Reserve (NGR): protected site on the left (western) side, and invaded areas used for agriculture on the right (eastern) side.

The proximity to water has always been an important factor that governs the distribution of rural communities. The human population that is directly or indirectly dependent on the PRF has grown over the past 40 years from 30,000 to a projected 400,000 in 2020 [5, 6]. The majority of the people dwelling the floodplain live in very poor conditions, being directly dependent on the floodplain. The water is used for both humans and livestock existence as well as for other goods from the floodplain such as fish (**Figure 2D**), fruits, reeds, thatch grass and firewood. The indirect dependence on the PRF is shown by the increase of agricultural lands that were once covered by natural vegetation (**Figure 3**). Practices of subsistence agriculture in the PRF introduced fertilisers and chemicals into the area, decreasing water quality and aquatic biodiversity. This together with alterations in the natural flow regime that was brought about by the construction of the dam, including irregular flood releases, prolonged droughts, deteriorating water quality, and organic pollutants, all severely threaten the functioning and biodiversity of the PRF.

Therefore, the aim of this chapter is to provide information on the current status of the freshwater diversity of the PRF comprising invertebrates, fishes, frogs and their parasitic fauna as well as their biological and physical stressors. Furthermore, the challenges and a way forward in managing the aquatic biodiversity of the PRF and proposed research to inform management and conservation strategies will be addressed herein. This chapter is based on the results of a decade (2010–2020) of research on the PRF led by the North-West University's Water Research Group (WRG).

2. Biodiversity of aquatic organisms from the lower Phongolo River and associated floodplains

2.1 Diversity of aquatic invertebrates

Aquatic invertebrates are invertebrates that require aquatic habitats to complete either one, several or all of their life stages in the aquatic environment [7]. Many terrestrial insects have a larval aquatic life stage including dragonflies and damselflies (Odonata), mosquitoes and flies (Diptera) and mayflies (Ephemeroptera) [8–10] and are therefore considered aquatic invertebrates in their larval phase. Truly aquatic or 'permanent' aquatic invertebrates complete all of their life stages in the aquatic environment and include zooplankton such as water fleas (Cladocera), micro-crustaceans (Copepoda) and fairy shrimp (Anostraca) as well as larger macroinvertebrates such as freshwater snails (Mollusca) and crabs (Potamonautidae)

[10–12]. Aquatic invertebrates are present in most freshwater ecosystems and respond rapidly to a broad range of physical and chemical environmental conditions such as changes in habitat condition or water chemistry [7, 13]. They are also quite immobile, particularly those that complete all their life cycles in water, and are constantly in contact with both bottom sediments and the water column [14]. Aquatic invertebrates are also in the unique position to act as the transitional link between lower level producers (including diatoms and algae) that they feed on and higher level consumers (including larger invertebrates, fish and birds) that feed on them [15–17]. For these reasons, aquatic invertebrates are ideal indicators of change in the aquatic environment and ecosystem health, particularly in lotic habitats, and have been used as such in various ecological assessments (e.g. [13, 18, 19]).

Aquatic biodiversity research in the PRF has been undertaken since the late 1960s, but it was not until 2012 that aquatic invertebrates were also included in ecological assessments of the region (see [2, 20, 21]). As part of these assessments, aquatic invertebrates were collected from the PR as well as many floodplain and temporary pans, both within and outside NGR. Approximately 131 taxa of aquatic invertebrates from 70 families have been identified from the river, while 117 taxa from 49 families and 109 taxa from 54 families have been collected from the floodplain and temporary pans, respectively (e.g. **Figure 4**). Many of the aquatic invertebrates collected from the river and floodplain pans both within and outside NGR include sensitive biota [7] such as riffle beetles (Elmidae), brush-legged mayflies (Oligoneuriidae), caseless caddisflies (Philopotamidae) and spiny crawlers (Teloganodidae).

Lake Nyamithi forms an important part of the assessment of the aquatic invertebrates of the PRF. This lake is quite unique as it is a large naturally saline lake and the only permanent wetland-type ecosystem located within the NGR [3, 22]. Due to its permanence, this habitat acts as a refuge to many of the aquatic and (semi-)

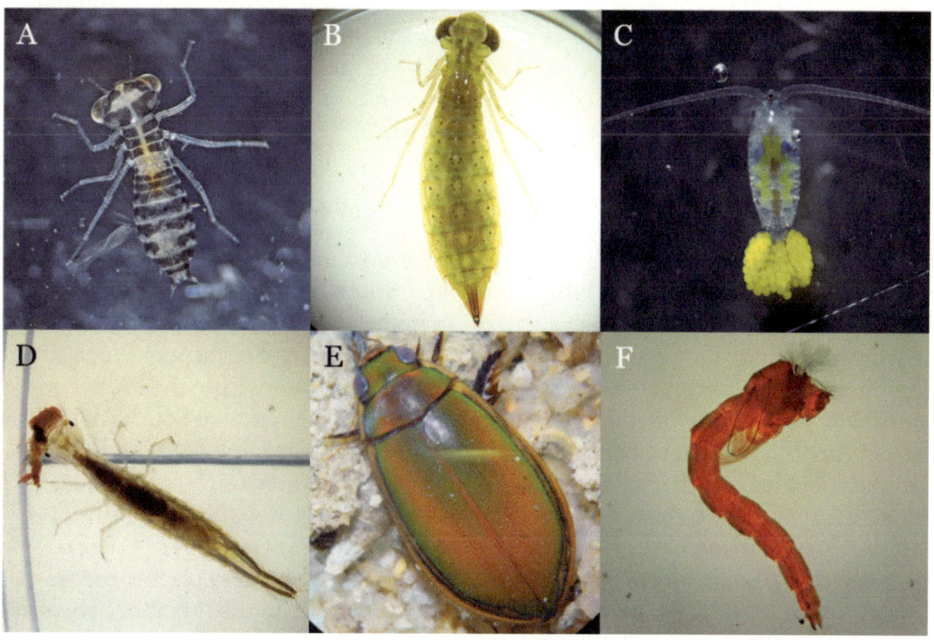

Figure 4.
Examples of aquatic invertebrates found in the lower Phongolo ecosystem. (A) Young dragonfly nymph (Aeshnidae); (B) mature dragonfly nymph (Aeshnidae); (C) female micro-crustacean with eggs (Cyclopoida); (D) larval predaceous diving beetle (Dytiscidae) feeding on a bloodworm (Chironomidae); (E) adult Dytiscidae beetle and (F) pupal stage of Chironomidae.

terrestrial biota of the region during the dry season when many of the pans dry out [2]. Even though it is naturally saline, the diversity and abundance of aquatic invertebrates found in this ecosystem (108 taxa from 47 families) is comparable to that found in the river and floodplain and temporary pans. Overall, the diversity of aquatic invertebrates in the PRF is greater than that of comparable lowland river (see [23, 24]), pan (see [25]) and saline lake (see [26]) ecosystems. These findings further demonstrate that the PRF is a biodiversity hotspot for aquatic invertebrates and of great ecological importance.

Worryingly, aquatic invertebrate assessments of this ecosystem have also indicated the possibility that the anthropogenic disturbances taking place outside NGR has a negative impact on the ecological integrity of the aquatic habitats of the PRF. Although biodiversity has generally been found to be greater in the river and pans located within NGR compared to outside the reserve, the river section that flows through the reserve is clearly not well protected given the types and abundances of certain aquatic invertebrates found. Several pollutant-tolerant taxa [7] such as the sharp-spired bladder snail, *Physa acuta*, and aquatic earthworms, Lumbriculidae, were found to be some of the most abundant taxa in the river inside NGR. Research found these types of taxa, particularly snails of the genus *Physa*, to flourish in systems that are nutrient enriched [27] suggesting that this may be the case for the PR. Additional stressors also include the invasive quilted melania snail, *Tarebia granifera*, which is distributed throughout the PR and is discussed further in Section 3.1.

Many of the pans outside the reserve are also threatened by anthropogenic stressors since they are utilised by local communities as a source of domestic and agricultural water and for subsistence fishing [2, 28, 29] all of which affects the integrity of pans. Ndumo Game Reserve therefore acts as a refuge for aquatic invertebrates of the PRF. This can be observed by the much higher number of invertebrate families that are present exclusively within NGR compared to outside (19 inside vs. 4 outside). Additionally, those invertebrate families present in high abundance in pans outside the reserve, namely rat-tailed maggots (Syrphidae), biting midges (Ceratopogonidae) and crane flies (Tipulidae) are known to be more tolerant of eutrophication and other forms of pollution [7] further demonstrating how these wetlands are being negatively affected.

2.2 Diversity of freshwater fishes

Fish have been regarded as one of the world's most important natural resources as they provide animal protein to billions of people annually, especially in developing countries [30]. Furthermore, fish also play an important role in freshwater ecosystem ecology where they can control prey such as zooplankton or be prey themselves for other reptiles, fish, mammals or birds. Fish communities within the PR and its associated floodplain wetlands are one of the most diverse found in South Africa [31] (see **Figure 5**). There have been 46 species recorded in this region from 12 different families (**Table 1**). Of these 46 species, *Oreochromis mossambicus* (Mozambique tilapia) (**Figure 5D**) is listed on the IUCN Redlist as a vulnerable species [40]. The South African Threatened or Protected Species List (TOPS) from 2013 [41] also included the *O. mossambicus* as a protected species in South Africa placing restrictions on its utilisation. This listing of *O. mossambicus* is due to the threat of hybridisation with the alien cichlid *Oreochromis niloticus* (Nile tilapia) [42] that could lead to a loss in genetic integrity. To date *O. niloticus* has not been recorded from the PRF thus making the *O. mossambicus* population from this region potentially one of only a few remaining genetically pure *O. mossambicus* populations left, although this needs confirmation through genetic studies. One of the unique fish species found in the PRF is the annual killifish, *Notobranchius orthonotus* (spotted killifish). This fish species is found in the temporary pans, specifically within NGR, as it is able to withstand drying through dormant eggs that are deposited in the sediment [43].

Figure 5.
Selection of fishes collected from the lower Phongolo River and floodplain: (A) Brycinus imberi *(max. length 19.8 cm); (B)* Hydrocynus vittatus *(max. Length 105 cm); (C)* Coptodon rendalli *(max. length 45 cm); (D)* Oreochromis mossambicus *(max. length 39 cm); (E)* Tilapia sparrmanii *(max. length 23.5 cm); (F)* Enteromius afrohamiltoni *(max. length 17.5 cm); (G)* Enteromius annectens *(max. length 7.5 cm); (H)* Enteromius paludinosus *(max. length 15 cm); (I)* Labeo rosae *(max. length 40 cm); (J)* Megalops cyprinoides *(common length 30–45 cm); (K)* Synodontis zambezensis *(max. length 43 cm); (L)* Petrocephalus wesselsi *(max. length 11.4 cm); (M)* Marcusenius macrolepidotus *(max. length 32 cm); (N)* Schilbe intermedius *(max. length 50 cm). Data on maximum length extracted from [32].*

Species	1974–1976	1983	1984	1993–1994	2012–2014	2016–2019
Alestidae						
Brycinus imberi (*B. lateralis*)	x	x	x	x	x	x
Hydrocynus vittatus	x	x	x	x	x	x
Micralestes acutidens	x	x	x	x	x	x
Anguillidae						
Anguilla bengalensis labiata (*A. nebulosa*)	x	x		x	x	x
Anguilla bicolor bicolor	x	x		x	x	
Anguilla marmorata	x	x		x	x	x
Anguilla mossambica	x		x	x	x	
Cichlidae						
Coptodon rendalli (*Tilapia rendalli*)	x	x	x	x	x	x
Oreochromis mossambicus	x	x	x	x	x	x
Oreochromis placidus					x	
Pseudocrenilabrus philander	x	x	x	x	x	x
Tilapia sparrmanii	x	x	x	x	x	x
Clariidae						
Clarias gariepinus	x	x	x	x	x	x
Clarias ngamensis	x	x	x	x		
Cyprinidae						
Enteromius afrohamiltoni	x	x	x	x	x	x
Enteromius annectens	x	x	x	x	x	x
Enteromius pallidus				x	x	
Enteromius paludinosus	x	x	x	x	x	x
Enteromius radiatus	x	x	x	x	x	x
Enteromius toppini	x	x	x	x	x	x
Enteromius trimaculatus	x	x	x	x	x	x
Enteromius unitaeniatus				x	x	
Enteromius viviparus	x	x	x	x		x
*Cyprinus carpio**				x	x	x
Labeo congoro	x	x	x	x		x
Labeo cylindricus	x	x	x	x	x	x
Labeo molybdinus	x	x	x	x	x	
Labeo rosae	x	x	x	x	x	x
Labeobarbus marequensis	x	x				x
Engraulicypris brevianalis	x	x	x	x		x
Opsaridium peringueyi (*zambezense*)				x		
Cyprinodontidae						
Nothobranchius orthonotus	x			x	x	x

Species	1974–1976	1983	1984	1993–1994	2012–2014	2016–2019
Gobiidae						
Awaous aeneofuscus				x	x	x
Glossogobius callidus				x	x	x
Glossogobius giuris	x	x	x	x	x	x
Redigobius dewaali	x	x	x	x	x	
Megalopidae						
Megalops cyprinoides				x	x	
Mochokidae						
Chiloglanis paratus	x		x	x	x	x
Chiloglanis swierstrai	x	x	x	x	x	x
Synodontis zambezensis	x	x	x	x	x	x
Mormyridae						
Marcusenius macrolepidotus	x	x	x	x	x	x
Petrocephalus wesselsi (*P. catostoma*)	x	x	x	x	x	x
Schilbeidae						
Schilbe intermedius	x	x	x	x	x	x
Sparidae						
Acanthopagrus berda						x
Syngnathidae						
Microphis fluviatilis				x		
Total no. of species	35	32	30	42	37	35

*Invasive species.

Table 1.
Comparative species list of the fishes collected from the lower Phongolo River and floodplain (data from [2, 33–39], and the present study).

Research on the fish community have been extensive since the 1970s during the construction of the Pongolapoort Dam. Fish surveys were completed in various decades from the 1970s to the most recent surveys from 2016 to 2019 (**Table 1**). The species diversity found from the 1970s to the current surveys show variation, potentially due to fluctuating rainfall, the flooding regime and management of flood releases from Pongolapoort Dam. Of the 46 expected species, only the intensive surveys of [34–39] came close to sample all of the expected species (42 species collected), highlighting how the species diversity varies depending on many different physical and biological aspects. Interestingly, a total of 43 species (one more than the number recorded during the intensive surveys in the 1990s) were recorded by the WRG research teams between 2012 and 2019.

The occurrence and diversity of fish in the floodplains are driven by their dependence on receiving enough water during the summer rainfall season. Often the dominance of cichlids (*O. mossambicus* and *Coptodon rendalli* [redbreast tilapia]) occur during periods of lower flows while a more equal distribution of species and biomass are present during higher flow periods [3, 33]. The success of the cichlids during lower flow periods are attributed to its resilience to survive in systems with higher electrical conductivity. Furthermore, of the various important fish species in the system, these are the only species able to spawn without increased flow and flow velocity triggering spawning events [3]. Smit et al. [2] indicated that two species [*Labeobarbus marequensis* (large-scale yellowfish) and *Engraulicypris brevianalis* (river sardine)] were potentially locally extinct, especially as the *Lb. marequensis* were not sampled during the 1990s and 2012–2014 surveys (**Table 1**). However, both these species were collected during the recent (2016–2019) surveys showing the resilience of the fishes and the system. There were eight species not collected during 2016–2019, which were collected during 2012–2014 (see **Table 1**), that might be the result of the extreme drought during the latter surveys. Furthermore, as no flood releases from the Pongolapoort Dam have been possible during this drought (since 2016), it is unlikely that these species are extinct from the PR, but should return from refugee areas in the Usuthu and Maputo Rivers once increased flows and flood releases occur.

The fish community in the PRF are important both ecological and economically. Firstly, many of the fish found in the PRF represent the southernmost distribution range of these species, making the system very important for biodiversity conservation. Secondly, the fish from the floodplain wetlands are extensively utilised (third most popular animal protein [44]) as a major source of protein by the local communities. Heeg and Breen [3] estimated that 400 tonnes of fish was harvested every year. The main fish species identified for consumption were *O. mossambicus* (**Figure 5D**), *C. rendalli* (**Figure 5C**), *Glossogobius giuris* (tank goby), and *Glossogobius callidus* (river goby). These species are also important fodder fish for *Hydrocynus vittatus* (tigerfish, **Figure 2B**), a popular recreational angling species in the PRF as well as in the Pongolapoort Dam [2, 45]. Thirdly, the fish communities are a crucial link in the food chain within the floodplain wetlands and especially in Lake Nyamithi where the abundances of fish serve as food sources to the unique and extremely diverse water bird community [2].

The 2012–2014 study [2] on the PRF investigated the Present Ecological State (PES), using the Fish Response Assessment Index. The results indicated that the fish community was in a seriously modified ecological state (Ecological Category of D/E); indicating an extensive or serious loss of natural habitat, biota and basic ecosystem functioning. This impacted state was especially evident downstream of the Pongolapoort Dam where high flow velocities are experienced during flood releases. The major threats to the fish communities were identified as physical impacts (unseasonal flood releases, poor water quality, water abstraction), over utilisation of fish from floodplains outside NGR, and biological threats (invasive species) [2].

2.3 Diversity of anurans

Amphibians are a diverse group that have adapted to a variety of habitats throughout the world. According to Frost [46], there are more than 8,203 known amphibian species globally, with more than 178 species in southern Africa. In recent decades, amphibians have suffered sudden, high mortality rates with many species becoming extinct [47, 48]. Although many factors contribute to this decline in amphibian populations, the most significant cause is habitat loss and fragmentation through anthropogenic disturbances. Paradoxically, the realisation that frog numbers are in decline sparked a renewed interest in amphibian biodiversity studies. The use of modern molecular and bioacoustic techniques and increased scientific surveys in remote areas, have resulted in new species being discovered and described frequently. Since 1985, the total number of globally recognised species has increased by over 60% [48–51].

In terms of amphibian diversity at a global scale, South Africa is currently ranked as the 27th country with the highest known species richness [52]. Within South Africa, the highest amphibian diversity can be found in the eastern part of the country and in particular, the province of KZN [53]. This province is an essential refuge for several endangered and endemic species. However, environmental stress, due to anthropogenic activities, poses a severe threat to the survival of amphibians, with safe havens becoming more critical for the conservation of amphibian species richness in South Africa. Currently, there are 71 different anuran species (excluding subspecies) in KZN; these account for 40% of the total frog diversity that occur in southern Africa [50, 54]. Northern KZN or Zululand has two humid, subtropical regions, namely, the Maputaland and the KZN coastal area. These areas are transition zones between tropical and temperate climates, characterised by high anuran species richness (see **Figure 6**) [52, 53, 55, 56].

Figure 6.
Frogs of the lower Phongolo river and floodplain. (A) Arthroleptis stenodactylus*; (B)* Leptopelis mossambicus*; (C)* Breviceps adspersus*; (D)* B. mossambicus*; (E)* Poyntonophrynus fenoulheti*; (F)* Schismaderma carens*; (G)* Sclerophrys garmani*; (H)* Scl. gutturalis*; (I)* Scl. pusilla *(J)* Hemisus marmoratus*; (K)* Afrixalus aureus*; (L)* Afr. delicates*; (M)* Afr. fornasini*; (N)* Hyperolius argus*; (O)* Hyp. marmoratus*; (P)* Hyperolius poweri*; (Q)* Hyp. pussilus*; (R)* Hyp. tuberilinguis*; (S)* Kassina senegalensis*; (T)* Phlyctimantis maculatus*; (U)* Phrynomantis bifasciatus*; (V)* Phrynobatrachus acridoides*; (W)* Phry. mababiensis*; (X)* Phry. natalensis*; (Y)* Xenopus laevis*; (Z)* X. muelleri*; (AA)* Hildebrandtia ornata*; (BB)* Ptychadena anchietae*; (CC)* Pty. mossambica*; (DD)* Pty. nilotica*; (EE)* Pty. oxyrynchus*; (FF)* Pty. porosissima*; (GG)* Amietia delalandii*; (HH)* Cacosternum boetgeri*; (II)* Pyxicephalus edulis*; (JJ)* Strongylopus fasciatus*; (KK)* Tomopterna adiastola*; (LL)* T. krugerensis*; (MM)* T. natalensis *and (NN)* Chiromantis xerampelina*. Photos not to scale.*

According to Du Preez and Carruthers [50], frogs inhabit almost every environment and habitat on the subcontinent. However, even though frogs are found in a variety of habitats, most species are specific to the particular habitat in which they are able to survive and more importantly, reproduce. Because of this specialisation many frog species are vulnerable to changing habitats and are directly affected by environmental disturbances [57].

The PRF is an area recognised as a biodiversity hotspot for amphibians, offering a great diversity of habitats suitable for amphibians ranging from big rivers, streams, pans, pools, swamps, marshland, rain-filled depressions and terrestrial habitats [58]. Currently, this selection of habitats caters for the 41 species known from the area (**Figure 6**) [59]. Furthermore, as a result of the survey work conducted during the current project the previously unknown Ndumo rain frog (*Breviceps passmorei*) (**Figure 7**) was discovered and described, the common name referring to the type locality [60].

Amphibians are a sensitive group, especially to rapidly changing environments, with many species only adapted to survive in specific habitat types. To gain a better perspective of how additional stressors such as habitat loss and fragmentation affect amphibians within the PRF in comparison to historical data, the current study undertook an extensive survey using both active and passive sampling techniques. While most anuran species are nocturnal, some species have prolonged breeding seasons whereas others are explosive breeders and are only active following rainfall events. Due to the unpredictability of weather and breeding activity, certain species are often overlooked during traditional active biodiversity surveys. Furthermore, each frog species has unique vocalisation calls serving as a valuable identification aid. In the current project, passive acoustic monitoring (PAM) was utilised in the NGR, with automated recorders were set up at two selected wetlands and set to record from 18:00 till 5:00 the following morning for 13 months. Findings, in combination with active sampling, were used to monitor biodiversity and breeding activity of frog species associated with the selected endorheic habitats.

In the current study, 83% (34/41) of the expected frog species were recorded based on the 75 years (1929–2004) of historical data. In the NGR alone, a total of 32 frog species were recorded, stressing the importance and value of natural protected areas and how they support not only specific species but whole communities. These results indicate that even though there are significant global amphibian declines, areas such as the NGR still provide a haven and refugia for frog species to flourish in, and should remain protected at all costs.

Figure 7.
*The Ndumo rain frog (*Breviceps passmorei*).*

Results from PAM indicated that the peak breeding season for the majority of the species, with 79% (15/19) calling males recorded, was in the southern hemisphere summer between December 2013 and January 2014. Of the 19 species of frogs recorded the hourly calling activity and intensity differed among species, with only four species, namely, *Sclerophrys pusilla*, *Phlyctimantis maculatus*, *Phrynobatrachus mababiensis* and *Phrynomantis bifasciatus* were recorded reaching an average call intensity of 5/5 (**Figure 8**). These findings are indicative of explosive breeders, with a high calling intensity for only a few weeks a year in correlation with rainfall patterns. Six species were recorded calling in all 12-h slots, namely, *Chiromantis xerampelina*, *Hemisus marmoratus*, *Hyperolius marmoratus*, *Phly. maculatus*, *Kassina senegalensis* and *Phry. mababiensis*. Furthermore, four species (*Afrixalus aureus*, *Afr. delicatus*, *Hyp. marmoratus* and *K. senegalensis*) call intensity and thus breeding activity peaked between 18:00 and 00:00, and for six species, (*Scl. pusilla*, *Phry. mababiensis*, *Phry. natalensis*, *Phryn. bifasciatus*, *Ptychadena anchietae* and *Pty. mossambica*) call intensity peaked between 00:00 and 5:00 (**Figure 8**).

As mentioned previously, the PRF area experienced a massive influx of people over the past four decades that resulted in a transformation of the landscape and fragmentation of natural habitats. These alterations place enormous pressures on the environment. Furthermore, the introduction of the invasive redclaw crayfish (*Cherax quadricarinatus*) (see Section 3.1) poses a severe risk to tadpole population density and should be considered high priority threat to biodiversity. However, the

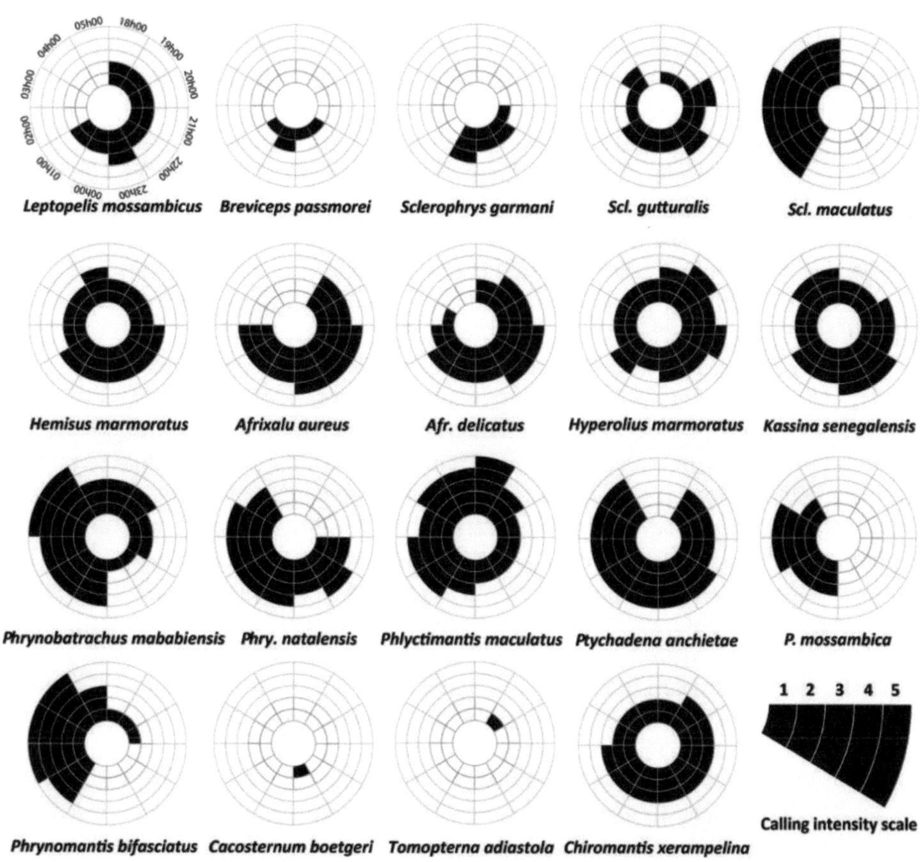

Figure 8.
The average hourly activity and intensity of male frog species. Displayed are the data on the average calling activity and intensity of each frog species recorded by the song metre for the first 10 min of each hour (18:00 to and including 05:00). The scale bar represents the hourly calling intensity.

use of educational posters and the field guide to the frogs of Zululand [59] helped inform and educate local communities on the importance of conserving the environment future generations.

2.4 Diversity of parasites of aquatic organisms

Parasites constitute a fundamental component of the global biodiversity, accounting for one third of the species on Earth, playing a key role in ecosystem functioning by regulating host population density and abundance, and being an integral part of food webs. However, parasites are still a neglected component in biodiversity surveys [61, 62]. Parasite inventories can provide knowledge to the comprehension of life cycles, pathological impacts, and evolution of host–parasite systems in aquatic environment [63].

Ectoparasites live on the surface of hosts, while endoparasites live internally [64]. The life cycle of parasites can be direct (monoxenous), requiring only one host to be completed, or indirect (heteroxenous), requiring one or more intermediate hosts to be completed [65]. A definitive (or final) host is the one in which the parasite is sexually mature; while the intermediate host is the one required for the development of parasitic stages without reaching sexual maturity. Paratenic or transport hosts are the ones in which parasites do not develop further but remains alive and infective [64]. Parasites exhibit variable degrees of host specificity, with some infecting only a single species or related species (specialists), and some infecting many unrelated species (generalists) [66].

A myriad of metazoan parasitic groups is found in aquatic organisms, including Protista, Myxozoa, Platyhelminthes [Monogenea, Trematoda (subclasses Aspidogastria and Digenea), Cestoda], Acanthocephala, Nematoda and Crustacea. Some species infecting aquatic organisms present zoonotic potential, such as certain digeneans, cestodes and nematodes. All parasitic groups have species that can harm aquatic hosts, especially when present at high intensity of infections that normally occurs in aquaculture scenarios or in the case of invasive parasites.

Parasites that naturally occur in wild fish normally do not cause negative impacts on host populations. The opposite is usually true as a high diversity of parasites in freshwater fishes in natural conditions are used as indicators of a healthy and functional ecosystem [67]. Moreover, parasites are very sensitive to environmental alterations such as pollutants, providing relevant information about the quality of a given system. Some parasites have been identified as sentinels for chemical pollution because these organisms can accumulate chemicals at a higher level compared to their hosts [68, 69]. Monogeneans ectoparasites are in direct contact with the environment, thus more sensitive to changes in water parameters. Studies on monogeneans of fishes as biomarkers have been conducted under different approaches, such as effects of high concentrations of effluents, hydrological cycle of floodplain areas, heavy metal concentrations, trophic concentrations of reservoirs, and prevalence and abundance in lotic and lentic environments of a river basin ([70] and references therein).

Research on freshwater fish parasites in the PRF is relatively new and still scant. Before the present study the only fish parasites recorded from the PR was two monogenean species by Price et al. [71] from the gills of *Tilapia sparrmanii* and *Enteromius trimaculatus*, respectively (see [72]). Svitin et al. [73] studied the diversity of camallanid nematodes from two catfishes, providing novel information on their morphology and genetic data; Hoogendoorn et al. [74] studied the diversity of digenean metacercariae (Diplostomidae) and also provided novel information on the morphology and genetic data of these parasites from freshwater fishes; Smit et al. [75] recorded for the first time trypanosomes in freshwater fish and in their leech vectors from this region, providing morphological and molecular characterisation of the haemoparasites; and recently, Schaeffner et al. [76] described a new cestode species from *S. zambezensis*. Information on the abovementioned records are shown in **Table 2** and some of the fish parasites are shown in **Figure 9**.

Parasite species	Locality	Host species	Reference
Trypanosomatida			
Trypanosoma mukasai	PR (NGR)	*Clarias gariepinus*, *Coptodon rendalli*, *Oreochromis mossambicus*, *Synodontis zambezensis*	[75]
Monogenea			
Characidotrema auritum	PR	*Brycinus imberi*	[77]
Pseudodactylogyrus anguillae	PR (NGR)	*Anguilla marmorata*	[78]
Cichlidogyrus papernastrema	PR	*Tilapia sparrmanii*	[71]
Dactylogyrus myersi	PR,	*Enteromius trimaculatus*	[71]
Macrogyrodactylus clarii	KP (NGR),	*C. gariepinus*	[79]
Macrogyrodactylus congolensis	UR (NGR)	*C. gariepinus*	[79]
Macrogyrodactylus karibae	KP, UR (NGR)	*C. gariepinus*	[79]
Quadriacanthus sp. 1	LN, KP, PR (NGR)	*C. gariepinus*	Present study
Quadriacanthus sp. 2	LN, KP, PR (NGR)	*C. gariepinus*	Present study
Digenea			
Diplostomum sp.	PR (NGR)	*S. zambezensis*	[74]
Diplostomum sp. 14	PR (NGR)	*Anguilla bengalensis labiata*, *O. mossambicus*, *S. zambezensis*	[74]
Cestoda			
Barsonella lafoni	LN	*C. gariepinus*	Present study
Tetracampos ciliotheca	LN, PR	*C. gariepinus*	Present study
Wenyonia gracilis	PR (NGR)	*S. zambezensis*	[76]
Nematoda			
Paracamallanus cyathopharynx	NL	*C. gariepinus*	[73]
Procamallanus pseudolaeviconchus	NL	*C. gariepinus*	[73]
Spirocamallanus daleneae	PR	*S. zambezensis*	[73]
Hirudinea			
Batracobdelloides tricarinata	PR (NGR)	*C. gariepinus*, *S. zambezensis*	[75]

Abbreviations: KP—KuShokwe Pan; LN—Lake Nyamithi; NGR—Ndumo Game Reserve; PR—Phongolo River; UR—Usuthu River.

Table 2.
Diversity of freshwater fish parasites naturally occurring in the lower Phongolo River and floodplain.

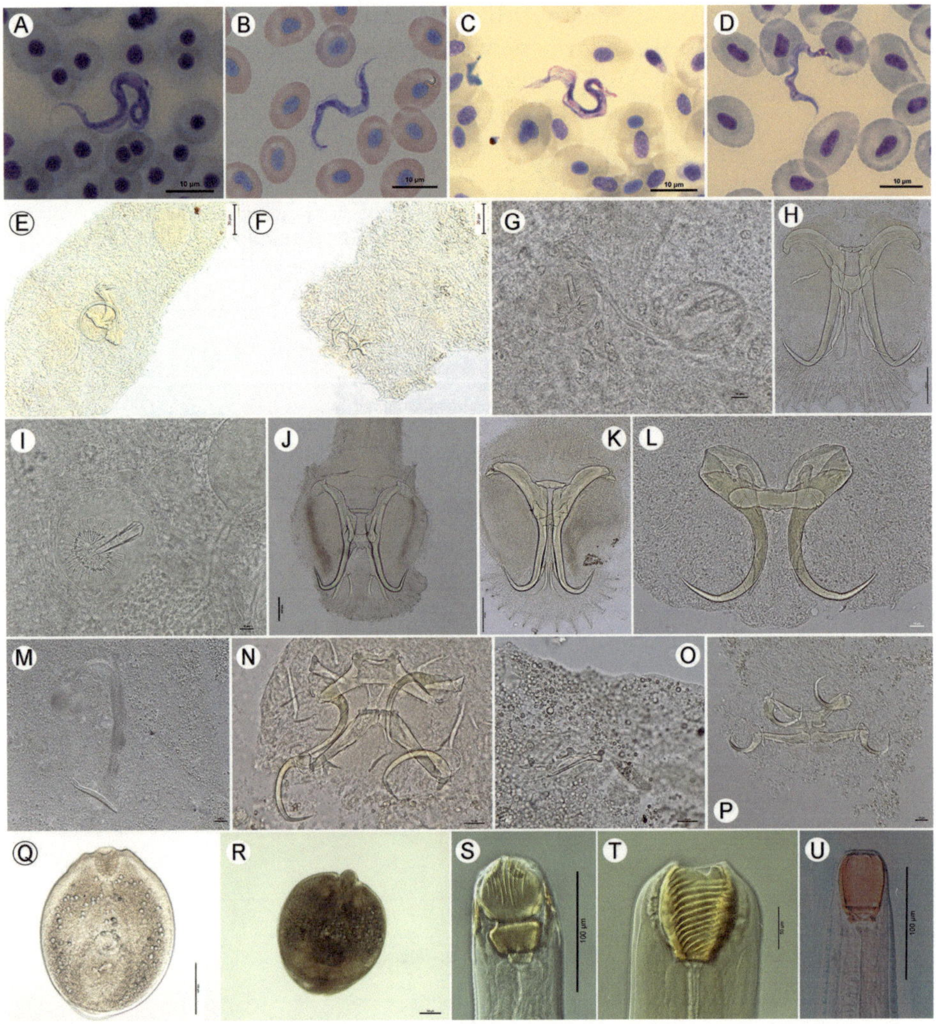

Figure 9.
Photomicrographs of some helminth parasites from freshwater fishes from the Phongolo River and floodplain. Trypanosoma mukasai *from: (A)* Clarias gariepinus; *(B)* Coptodon rendalli; *(C)* Oreochoromis mossambicus; *(D)* Synodontis zambezensis. Characidotrema auritum *(Monogenea: Dactylogyridae) from* Brycinus imberi: *(E) male copulatory complex (MCO); (F) haptor.* Macrogyrodactylus clarii *from* C. gariepinus: *(G) MCO; (H) haptor.* Macrogyrodactylus congolensis *from* C. gariepinus: *(I) MCO and (J) haptor.* Macrogyrodactylus karibae *from* C. gariepinus: *(K) haptor.* Pseudodactylogyrus anguillae *from* Anguilla marmorata: *(L) haptor; (M) MCO.* Quadriacanthus *sp. 1 from* C. gariepinus: *(N) haptor; (O) MCO.* Quadriacanthus *sp. 2 from* C. gariepinus: *(P) haptor.* Diplostomum *sp. 14 from* O. mossambicus: *(Q) total view.* Diplostomum *sp. from* S. zambezensis: *(R) total view.* Paracamallanus cyathopharynx *from* C. gariepinus: *(S) anterior end with detail of buccal capsule.* Spirocamallanus daleneae *from* S. zambezensis: *(T) anterior end with detail of buccal capsule.* Procamallanus pseudolaeviconchus *from* C. gariepinus: *(U) anterior end with detail of buccal capsule.*

Amphibians are well suited hosts for parasites, as most parasites rely on aquatic infective stages for transmission and reproduction, much in the same way that amphibians depend on extended exposure to aquatic systems for reproduction and their life history. This in turn increases the host–parasite contact rates. Frogs serve as host for most parasite groups including protozoans, monogeneans, digeneans, cestodes, acanthocephalans, nematodes and mites [50]. Furthermore, frogs are often infected with metacercaria where the frog serve as an intermediate host for a definitive reptile or bird host. However, parasites of frogs and other ectotherms have co-evolved over a long period of time and although these parasites are still true parasites per definition, they seldom inflict adverse effects on their hosts, as compared to endotherm hosts [80, 81]. Parasites that have thus far been encountered in frogs from the PRF are presented in **Table 3**. Additionally, parasites reflect their host species' environmental interactions, revealing feeding behaviour, geographical ranges and social systems [87]. Blood parasites present good examples of these interaction based on their various transmission strategies, such as via consumption of an infected intermediate host or directly from a blood feeding vector. Frogs from the PRF were shown to harbour a number of protozoan blood parasites ranging from intracellular apicomplexan parasites to extracellular euglenozoan flagellates (see **Figure 10**) [85, 88].

Considering the diversity of fishes and amphibians of the lower PRF, future studies on the diversity of their parasites can reveal unique ecological characteristics of the environment, more insights on the drivers of host–parasite relationships like

Parasite species	Locality	Host species	Reference
Protista			
Dactylosoma kermiti	PR (NGR)	*Ptychadena anchietae*	[82]
Hepatozoon involucrum	PR	*Hyperolius marmoratus*	[83]
Hepatozoon ixoxo	PR (NGR)	*Hemisus marmoratus*, *Sclerophrys garmani*, *Sclerophrys gutturalis*, *Sclerophrys pusilla*, *Ptychadena mossambica*, *Ptychadena nilotica*	[84, 85]
Hepatozoon tenuis	PR	*Afrixalus fornasini*, *Hyperolius argus*, *Hyp. marmoratus*	[83]
Hepatozoon thori	PR	*Hyp. argus*, *Hyp. marmoratus*	[83]
Hepatozoon sp.	PR (NGR)	*Ptychadena anchietae*,	[85]
Haemococcidia sp.	PR (NGR)	*Pty. anchietae*, *Phrynobatrachus mababiensis*	[86]
Trypanosoma spp.	PR (NGR)	*Afr. fornasini*, *Hem. marmoratus*, *Hyp. argus*, *Hyp. marmoratus*, *Hyperolius tuberilinguis*, *Pty. anchietae*, *Pty. mossambica*, *Scl. gutturalis*, *Scl. pusilla*	[85]
Monogenea			
Polystoma vernoni	NGR	*Ptychadena oxyrhynchus*	Present study
Protopolystoma orientalis	NGR	*Xenopus muelleri*	Present study
Cestoda			
Cephaloclamys sp	NGR	*Xenopus muelleri*	Present study
Nematoda			
Cosmocerca sp.	NGR	*Kassina senegalensis*	Present study
Cosmocerca sp.	NGR	*Ptychadena anchietae*	Present study
Cosmocerca sp.	NGR	*Tomopterna tandyi*	Present study
Aplectana sp.	NGR	*Breviceps passmorei*	Present study
Camallanus kaapstaadi	PR (NGR)	*Xenopus mulleri*	Present study
Batrachocamallanus xenopodis	PR (NGR)	*Xenopus mulleri*	[73]

Abbreviations: NGR—Ndumo Game Reserve; PR—Phongolo River.

Table 3.
Diversity of parasites occurring in frogs in the lower Phongolo River and floodplain.

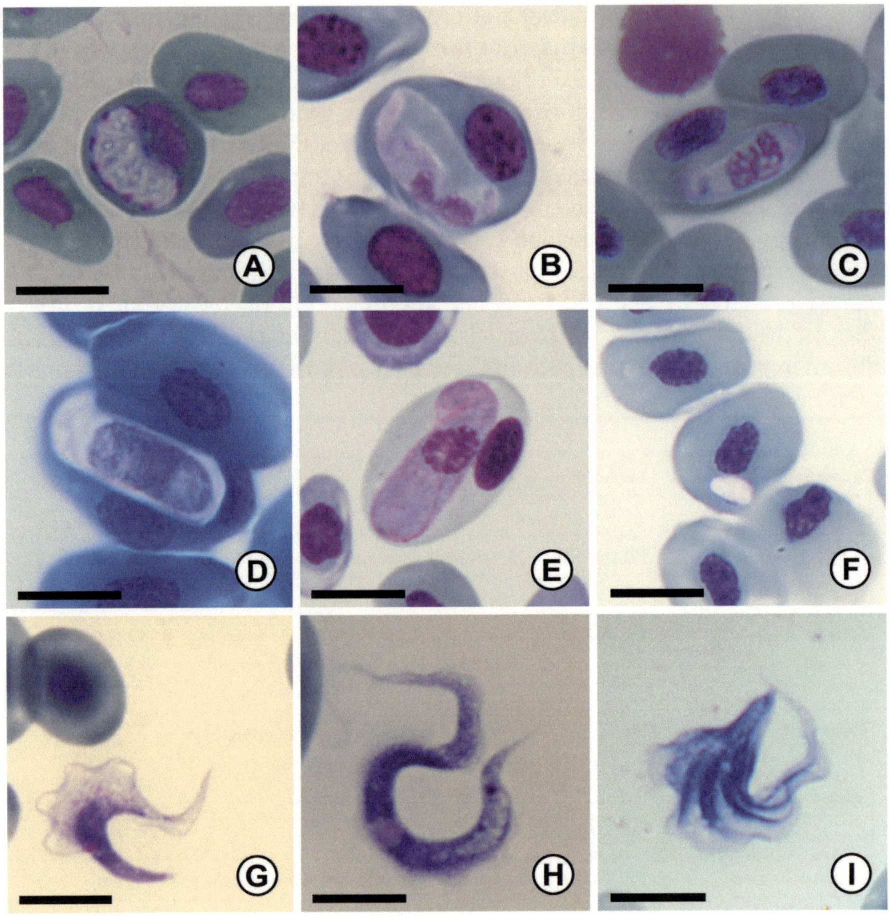

Figure 10.
Photomicrographs of various frog blood parasites encountered in the Phongolo River and floodplain. (A) Dactylosoma kermiti *infecting* Ptychadena anchietae*; (B)* Hepatozoon involucrum *found infecting* Hyperolius marmoratus*; (C)* Hepatozoon ixoxo *infecting* Sclerophrys gutturalis*; (D)* Hepatozoon thori *infecting* Hyperolius argus*; (E)* Hepatozoon *sp. infecting* Pty. anchietae*; (F)* Haemococcidia *sp. infecting* Ptychadena anchietae*; (G)* Trypanosoma *sp. infecting* Pty. anchietae*; (H)* Trypanosoma *sp. infecting* Scl. gutturalis *and (I)* Trypanosoma *sp. infecting* Sclerophrys pusilla*. Scale bar 10 μm.*

patterns of host-specificity specially for monogeneans, as well as unravel parasite species that are still new to science, thus increasing the knowledge about the aquatic biodiversity of South Africa.

3. Biological and physical stressors of aquatic organisms from the lower Phongolo River and associated floodplains

3.1 Invasive and alien aquatic species

Aquatic organisms are intentionally moved and introduced through several pathways and occasionally their introduction is accidental. Natural introductions of alien species occur when dispersed beyond its natural distribution range with connectivity of two geographic regions during floods (pans, rivers and streams) or migration of species that host parasitic organisms. Human mediated introductions are more common and deliberate when introduced for breeding, ornamental and recreational purposes (angling) or as a food source (aquaculture). Meanwhile, accidental introductions arise with escapes from captivity (aquaculture), uncontrolled releases of angling and ornamental species, inter-basin water transfer schemes, underlying contaminants or as stowaways on vessels and aquaria or farmed species [89–91].

Contaminants, pathogens or parasites accompanying introduced host species such as aquatic plants, molluscs, crustaceans and fish species are only noticed with increased infection, disease or mortality of the associated hosts. Often when an alien host is introduced it loses some of its native parasites (enemy release), however alien species may acquire native parasites and amplify transmission of pathogens or parasites of the native hosts (spillback). Alien host species can also 'dilute' transmission and infection of native pathogens and parasites (see [92, 93]). In contrast to this, parasites and pathogens that are co-introduced only infect the alien host and is co-invasive once it spills over to native biota (see [94] and references therein, [95]). Once alien species are introduced and established in the novel environment, they can become pests or threaten native biota and are then invasive or co-invasive (for parasites).

Alien or invasive species can impact aquatic ecosystem services and biological diversity through predation, parasitism, competition, hybridisation, habitat use and food web alterations [96, 97]. Livelihoods dependent on freshwater ecosystem and its resources can also be affected if aquatic diseases arise because of alien or invasive species, influencing marketability and commercial value for species used as a staple by local communities.

To date, eight invasive species have been recorded from the PRF (see **Table 4**). The first invasive to be recorded in the lower Phongolo was the common carp *Cyprinus carpio* in 1993 [34]. Apart from competing for resources, impact on water quality and threat to larval populations of macro-invertebrates, *C. carpio* is a known host to a variety of co-invasive parasites. The widespread Asian tapeworm *Schyzocotyle acheilognathi* and the anchor worm *Lernaea cyprinacea* has been co-introduced into the PRF with *C. carpio* and has spilled over to two native small barb species *Enteromius annectens* (broad-stripped barb) and *Enteromius bifrenatus* (hyphen barb), and to the vulnerable native *O. mossambicus*, respectively (see **Table 4**). The presence of these two co-invaders is concerning as infections with *S. acheilognathi* is associated with high infection intensities and pathology of the gut lumen, while *L. cyprinacea* cause haemorrhagic ulcers on the body surface of fish, leading to increased susceptibility to secondary infections [100 and references therein]. A recent study in the lower Phongolo also confirmed that the overall health state of fishes can be compromised when heavy infestations with *L. cyprinacea* occur [102]. This is of concern since *O. mossambicus* is a vulnerable native species and plays important economic and ecological roles in the PRF (see Section 2.2).

Furthermore, three invasive freshwater snail species, a crayfish and its co-invasive parasite inhabits the waters of the lower Phongolo. The reticulate pond snail *Lymnaea columella*, *P. acuta* and *T. granifera* directly competes with and affect the distribution of three native freshwater snails species: the common pond snail *Lymnaea natalensis*, *Bulinus africanus* and the red-rimmed melania *Melanoides tuberculata* [2, 101, 103]. A single niche competitor, the invasive *C. quadricarinatus*, which escaped from aquaculture farms in Swaziland, was first recorded from the PRF in 2013 [99]. In addition to posing a threat through direct predation on tadpoles, habitat modification and the resident native Natal freshwater crab *Potamonautes sidneyi*, the *C. quadricarinatus* in the PRF also hosts a co-invasive temnocephalan *Diceratocephala boschmai* (see **Table 4**) [99]. To date, no spillover of this temnocephalan to native freshwater crab, shrimp and other branchiopods have been noted. However, it has been experimentally proven that one of three co-invasive temnocephalan species infecting invasive freshwater crayfish species in South Africa can utilise native freshwater crabs as a host [91]. Some of the invasive organisms mentioned are shown in **Figure 11**.

Species	Vector/pathway into RSA	Presence	Infection of native species	Reference
Diceratocephala boschmai (Platyhelminthes: Temnocephalida: Diceratocephalidae)	*Cherax quadriacanthus* (crayfish) escapes from aquaculture farm in Swaziland [98]	LN	No record of spread to native freshwater crabs, shrimps and other Branchiopoda.	[99]
Schyzocotyle acheilognathi (Cestoda: Bothriocephalidae)	*Cyprinus carpio* (common carp)	LN, PR, P	*Enteromius annectens*; *Enteromius bifrenatus*	[100]
Lymnaea columella (Gastropoda: Lymnaeidae)	Stowaway on aquarium plants		–	[2]
Physa acuta (Gastropoda: Physidae)	Stowaway on aquarium plants	LN, PR	–	[2, 101]
Tarebia granifera (Gastropoda: Thiarinae)	Stowaway on aquarium plants	LN, PR	–	[2, 101]
Lernaea cyprinacea (Arthropoda: Lernaeidae)	*Cyprinus carpio*	LN, PR	*Oreochromis mossambicus*; *Coptodon rendalli*	[2, 102]
Cherax quadricarinatus (Arthropoda: Parastacidae)	Escape from aquaculture farm in Swaziland [98]	LN, PR, UR	–	[99]
Cyprinus carpio (Cypriniformes: Cyprinidae)	Recreational angling	Various sites in lower PR	–	[34]

Abbreviations: LN—Lake Nyamithi; PR—Phongolo River; P—Pumphouse; UR—Usuthu River.

Table 4.
Introduced aquatic species present in the lower Phongolo system.

Figure 11.
Invasive species in the lower Phongolo River and floodplain: (A) Cyprinus carpio *(Cypriformes: Cyprinidae) (common length 31 cm); (B)* Tarebia granifera *(Gastropoda: Thiarinae); (C)* Cherax quadricarinatus *(Arthropoda: Parastacidae) (max. Length 35 cm) from Ndumo game reserve; (D)* Diceratocephala boschmai *(Platyhelminthes: Temnocephalida: Diceratocephalidae) (length 832–1500 μm) parasite from* C. quadricarinatus. *Information on length extracted from [32, 104, 105], respectively.*

3.2 Physical and chemical stressors

The construction of the Pongolapoort Dam (Jozini Dam) in the early 1970s is arguably the greatest agent of change in the PRF. Following the construction of the dam, the downstream terrestrial physical and chemical template of the system was irrecoverably altered with a complicated interaction between the benefits and threats derived from the damming of the PR. Prior to the construction of the dam, the subsistence fisheries and agriculture practices were synchronised with the natural floods and their flooding and alluvial sediment deposit regimes of the floodplain pans [3]. Since its construction, there have been various 'flood release' strategies to simulate the natural floods in the floodplain with varying degrees of success [4]. Recently, Brown et al. [106] developed a set of holistic environmental flow recommendations that would supposedly meet the ecosystem and agricultural water requirements. However, these have yet to be evaluated since the region has been in the grip of a long-term drought that has negated water releases, other than base flows of 8 $m^3 \cdot s^{-1}$, since 2015 to the present [1].

As mentioned previously, the assurance of water resources and potential for sustained larger-scale agriculture resulted in exponential growth in the human population that is directly or indirectly dependent on the PRF. These communities are considered to be among southern Africa's poorest and traditionally highly dependent on harvesting natural resources. For these impoverished communities, living on or near the PRF allows them to maintain subsistence agricultural activities, provide for livestock, water collection for household usage, religious activities, a source of protein and sustenance through fishing and lastly the harvesting of plants such as water lilies, reeds and thatching grass [2, 5].

The conversion of natural land cover to agricultural landscapes is one of the main causes of environmental degradation and is a known driver of pollution of surface waters and subsequent loss of habitats and biodiversity [107]. Between 1955 and 2003, 40% of the natural floodplain vegetation was transformed into agricultural land [5]. The inadequate accessibility to water leads to further exploitation of the limited water resources of the region. Recent years have seen a marked increase in informally installed pumps along the length of the river, mostly used to irrigate the fields of the subsistence farmers. The water systems of the floodplain are further contaminated by oil and fuel leaks from the aforementioned pumps and washing of clothes. Waste dumping is common practice in South Africa and is not solely limited to urban areas, as is evident throughout the rural communities surrounding the floodplain and this poses significant risks to human and environmental health. Such practices even created waste management issues within the NRG.

The excessive reliance on the floodplain and its resources could be considered unsustainable and thus leads to significant biodiversity losses. Less than 10% of the entire floodplain is formally conserved in the NGR. Outside of the reserve, the riparian forests and vegetation are removed, being the wood used as fuel and the cleared areas for subsistence agriculture (see **Figure 3**). In 2008, the communities along the eastern boundary took down the 11 km long fence to gain access to 1000 hectares of the reserve for resources and establishing subsistence agriculture. As such, the only intact riparian zone is a 3-km stretch along the western bank of the active channel of the PR within the reserve.

The altered water releases, drought and increased pesticide use in the floodplain has resulted in increased chemical stressors being released into surface and groundwater of the system. Historically, the floodplain has been subjected to fluctuations in drought periods and flood inundation. As a consequence, highly saline conditions (total dissolved solids in excess of 5000 $mg \cdot l^{-1}$) have been reported in the floodplain pans that are prone to seepage of salts from the underlying marine cretaceous geological structure [3]. The salinities subsequently decrease when the pans are flooded during the next flood event. Due to the prolonged drought from 2014 and the absence of flooding since 2016, there has been a steady increase in electrical conductivity and nutrients. However, these levels still seem very similar to the historical conductivity and nutrient concentrations [1]. It would thus seem that the system is highly resilient to altered flow conditions with large fluctuations in water quality.

Large-scale organic pesticide application has been in place on the PRF for many decades. Dichlorodiphenyltrichloroethane (DDT) has been used for malaria vector control since the 1930s [108]. Sereda and Meinhardt [109] studied the levels of pesticides in surface- and groundwater of water bodies of the PRF. They attributed the presence of pyrethroids (cypermethrin, λ-cyhalothrin and cyfluthrin), organophosphates (fenthion and fenitrothion) organochlorines (DDT and its metabolites—pp-DDD and pp-DDE) and carbamates (carbosulfan and carbofuran) to agricultural and malaria vector control applications. The first records of DDT in fish and other wildlife from the PRF were reported by [110]. Bouwman et al. [111] concluded that the high levels of DDT and its metabolites found in human serum and breast milk from the PRF could be attributed to a combination of consuming contaminated fish and direct exposure through spray drift.

Pesticide usage and subsequent spillage and run-off remain a major concern due to the numerous negative impacts on wildlife and ecosystems. Organochlorine pesticides (OCPs) are a group of pesticides which have been banned globally due to their persistence, attested by the quantifiable concentrations even though usage has been banned for over two decades, as well as consequent negative impacts on the environment [2, 112, 113]. Volschenk et al. [114] highlighted the renewed focus on in quantifying OCPs in aquatic ecosystems in and around the floodplain. These chemicals accumulate throughout the food web and represent a significant threat to the aquatic diversity of the floodplain. The floodplain and its surrounding area are still classified as a malaria endemic area and as such has consistently been sprayed with DDT for the purpose of malaria vector control. General pesticide use has also increased due to increasing agriculture activities. Recently, [114] recorded a range of pesticides, including discontinued pesticides such as lindane in five of the most economically and ecologically important fish species of the floodplain. Contamination levels in the floodplain were lower than other regions across South Africa. They are however a reason for concern due to the risks posed to both humans and animals reliant on the floodplain, including important protected and red data listed species such as the Nile crocodile (*Crocodylus niloticus*), pelicans (*Pelecanus onocrotalus*) and Saddle billed storks (*Ephippiorhynchus senegalensis*). As anticipated, DDT and its metabolites were the dominant OCPs and were shown to magnify through the food chain. DDTs are associated with several harmful effects on fish, crocodile and bird populations across the world, including but not limited to reproductive impacts, endocrine disruptions, eggshell thinning and ensuing population declines. Importantly, as OCPs do not adhere to any type of border, conservation efforts that are implemented within the region should strive towards better management and increased public education.

4. Conceptualising conservation approaches for the PRF

The previous sections explain the biodiversity and conservation importance of the PRF whilst also highlighting particular conservation challenges. In order to conserve this unique biodiversity, different conservation governance and/or management approaches exist. Therefore, this section briefly explains the different approaches available in South Africa, for consideration in the Phongolo context. Although we explain the different approaches, we do not prescribe any specific approach. The entire suite of options needs to be considered to inform the design of a combined or hybridised tapestry of approaches, best suited to achieve the conservation objectives of the PRF. There is therefore no silver bullet or single solution when it comes to identifying conservation governance and management approaches. We frame our brief discussion around approaches and instruments already developed within the South African context more generally (for a more detailed discussion of these approaches and instruments see [115]). Three broad approaches are distinguished as illustrated by the three circles in **Figure 12**, namely: command- and control (CaC)-based, fiscal-/market-based and civil-based approaches. In relation to each approach, different so-called governance and management instruments are identified. Some of these instruments are considered hybridised instruments nesting between more than one approach, illustrated by the overlapping areas between the three circles numbered A, B, C and D.

The CaC approach includes those management instruments provided for by legal means. This is the most basic or classical approach centred on the understanding that the best way to control human behaviour is to enact laws and then enforce them—or the 'stick approach'. In terms of conservation, South Africa has a complex legal framework (for a detailed discussion of the legal framework see [116, 117]) and a myriad of instruments covering strategic and project level decision making. Strategic level instruments inform decision making at a policy and planning level such as the National Biodiversity Framework, Bioregional Plans, Biodiversity Management Plans and the proclamation of different protected areas (PAs). An example of these with in the PRF, is the formally proclaimed NGR. At project level CaC-based instruments include the issuing of permits and/or prohibition notices for legally defined 'restrictive activities' in relation to listed ecosystems and species (threatened or alien invasive). The protection of species through CaC has obvious relevance to the wealth of biodiversity in the PRF, highlighted in previous sections. While the CaC arrangements are considered critical for any conservation regime, failures of CaC instruments are well documented in the literature [118, 119]. The strengths of CaC are that it provides a high level of certainty by defining for example boundaries of PAs and prescribing behaviour in relation to listed species and ecosystems. The weaknesses relate to the resources required and time it takes

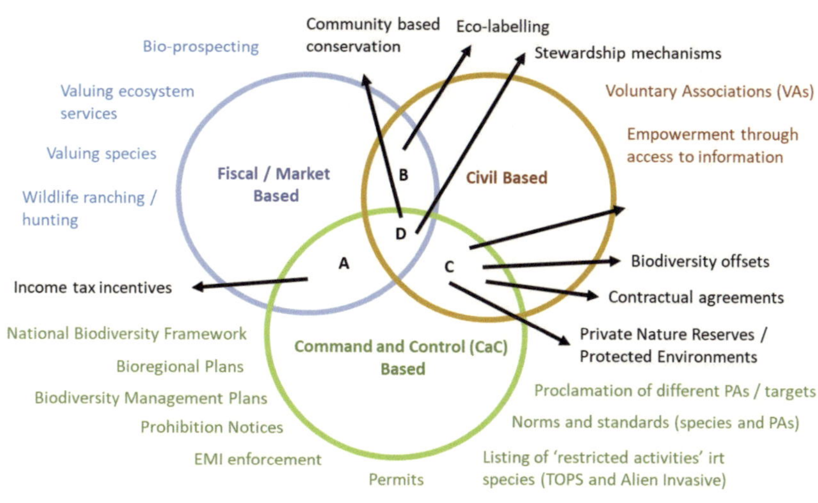

Figure 12.
Different conservation governance and management approaches within the south African context.

to ensure enforcement of these legal mechanisms. For this reason, additional approaches have emerged to supplement and support CaC arrangements such as fiscal- and civil-based approaches.

Fiscal-based approaches aim to use the market or price mechanism to change or incentivise certain behaviour, also sometimes referred to as the 'carrot approach'. In the case of conservation this refers to valuing ecosystem services and/or species within a market-based economy to incentivise their protection. The wildlife and bioprospecting industries are typically integral to this market-based system. The strengths of market-based instruments are the instant effect it seems to have on behaviour, unlike CaC-based instruments. However, pricing ecosystems and species have been controversial, especially in relation to the trade in endangered species – ivory and rhino horn a case in point. Market-based instruments are not well developed in the PRF and South African in general. This is mainly due to the significant inherent challenges in commodifying ecosystems and wildlife [120].

Civil-based instruments centre around empowerment of civil society in order to affect behaviour. The need for civil-based instruments derives from the acknowledgement that governments alone cannot deliver on conservation goals and objectives. Civil-based instruments include opportunities for civil society to organise themselves through for example voluntary associations (VAs) and empowerment of civil society through access to information on for example conservation-related matters. The PRF is an example of a highly complex civil society context with a range of civil society actors such as tribal authorities, farming unions, water user associations, etc. The advantage for the study area is that civil society seems well organised and represented. However, the challenge is to achieve a common understanding and general agreement on the future of conservation in the area amidst the range of actors.

In an ever-changing complex world, it is also evident that many of the most effective management instruments do not fit neatly into a specific approach but rather are designed as so-called hybrid instruments. For example, income tax incentives for landowners promoting conservation on their land is a clear hybrid between CaC and market-based approaches—see 'A' in **Figure 12**. This is because the tax incentive is incorporated into law, but the incentive is market based. We are not aware of any example within the PRF where this instrument has been used. Furthermore, instruments such as eco-labelling aims to make certain products more appealing to civil society or consumers based on their environmental performance—see 'B' in **Figure 12**. The hybrid approaches between civil- and CaC-based approaches shown as 'C' in **Figure 12** have a high level of potential relevance to the study area. These include biodiversity management agreements or contractual agreements between private or communal landowners and the state. The conservation status of private or communal land can also be formalised trough different kinds of protected areas such as Private Nature Reserves (PNRs) and Protected Environments [121]. An extension of the latter in recent years has been so-called community-based conservation (CBC) and conservation stewardship programmes that are considered the most integrated conservation instruments combining CaC, fiscal- and civil-based approaches—see 'D' in **Figure 12**. For example, the stewardship programme combines different levels of legal protection with money saving incentives and strong community involvement and ownership. Much research has been conducted on the success and failures of CBC generally, but also within the KZN Province, which could inform the future application within the study area (see [122–124]).

There are two guiding principles when considering these approaches, firstly the use of multiple approaches and instruments is preferred especially within a complex and multi-faceted context such as the PRF. Multiple approaches also enhance the redundancy effect by ensuring multiple possible solutions for a range of conservation challenges. Secondly, hybrid approaches are preferred (as represented by A, B, C, and D in **Figure 12**) that aim to optimise and merge the strengths of different approaches into single instruments. Ultimately the design and selection of conservation approaches for the PRF will depend on the agreed context specific conservation goals and objectives.

5. Conclusions

It is clear from the information presented in the preceding sections that South Africa's PRF is, in terms of aquatic organisms, highly diverse and of national and international importance. However, the aquatic ecosystem of the PRF is also under extreme pressure from a vast array of human activities, ranging from broadscale influences such as climate change and associated extreme weather events to local-scale impacts such as pollutants, over utilisation and invasive species. Interestingly, the PRF ecosystem has also exhibited tremendous resilience in dealing with all these anthropogenic stressors. Despite more than 10-fold increase in the human population depending on the ecosystem services provided by the PRF, a 5-year ongoing below-rainfall period with no flood release from the Pongolapoort Dam and reduction in the size of the area under formal protection, there is currently no evidence that points towards a loss in species diversity. However, at least for fishes, Smit et al. [2] showed that although all species are still present, there has been a clear shift in the community structures and dominance. The main question for the PRF therefore still remains; how long, and especially in the light of the continued lack of flooding events and increasing human settlement and activities, can the socio-ecological system of the PRF stay resilient before complete collapse?

To answer this very important question, we propose that future research into this and similar systems in Africa and globally should follow the One Health approach. The One Health approach deals with a multidisciplinary and collaborative approach to ensure optimum health of humans, animals and the environment. The importance of, and need for, this approach has really come to the front during the Covid-19 pandemic of 2020 where the world was brought to a standstill due to a virus that originated from an animal (zoonotic disease). Therefore, studies on specifically environmental health are urgently required in order to gain a better understanding of the causes and consequences of anthropogenic activities, and how these in return have an impact on human health. Future research should thus aim at investigating the possible natural hazards associated with effects of climate change (i.e. droughts and floods), in combination with environmental pollutants, on the severely threatened aquatic ecosystem of the PRF. Specific research objectives should include:

- Determine the risk and possible impact of climate change on water quantity and quality, and food security.

- Identify different water quality governance approaches and instruments to ensure environmental conservation and human health protection amidst a changing climate.

- Identify dimension of vulnerability and characteristics of resilience that make local communities more, or less, susceptible to water-related disaster risks as a result of climate change.

- Monitoring the introduction and spread of invasive aquatic animals (both free-living and parasitic).
- Develop and validate holistic environmental flows to maintain and support ecosystem structure and functions.
- Develop and apply a risk assessment framework that integrates the ecological, human and wildlife factors to evaluate the socio-ecological consequences of climate change-induced changes in water quality and quantity.

Although the abovementioned recommendations are specifically proposed for the PRF, these research questions, aims and objectives are applicable to all threatened floodplain systems globally.

Acknowledgements

The authors would like to express their gratitude to the following people for providing photomicrographs used in this chapter: Eva Řehulková (Faculty of Science, Masaryk University, Brno, Czechia); Tomáš Scholz, Blanka Škoríková, and Camila Pantoja (Institute of Parasitology, Biology Centre of the Czech Academy of Sciences, České Budějovice, Czechia); Coret Hoogendoorn and Roman Svitin (Water Research Group and African Amphibian Conservation Research Group, Unit for Environmental Sciences and Management, North-West University, South Africa) and Charles Griffiths (University of Cape Town, South Africa), and Andrew Wallis, CFD Productions, Potchefstroom. Research reported in this chapter benefitted from funding during the period from 2010 to 2020 from AIRES-Sud, a programme of the French Ministry of Foreign and European Affairs implemented by the Institut de Recherche pour le Développement (IRD-DSF); Water Research Commission of South Africa (Project K5-2185, PI: NJ Smit); VLIR-UOS TEAM project (ZEIN21013PR396, PI: L Brendonck and V Wepener) and the National Research Foundation (NRF) of South Africa (NRF project CPRR160429163437, grant 105979, NJ Smit, PI). Opinions expressed and conclusions arrived at are those of the authors and not necessarily those of the funding bodies.

Conflict of interest

The authors declare no conflict of interest.

Author details

Aline Angelina Acosta[1*], Edward C. Netherlands[2], Francois Retief[3], Lizaan de Necker[1,4], Louis du Preez[2,4], Marliese Truter[1,5], Reece Alberts[3], Ruan Gerber[1], Victor Wepener[1], Wynand Malherbe[1] and Nico J. Smit[1]

1 Water Research Group, Unit for Environmental Sciences and Management, North-West University, Potchefstroom, South Africa

2 African Amphibian Conservation Research Group, Unit for Environmental Sciences and Management, North-West University, Potchefstroom, South Africa

3 Unit for Environmental Sciences and Management, North-West University, Potchefstroom, South Africa

4 South African Institute for Aquatic Biodiversity (NRF-SAIAB), Makhanda, South Africa

5 DSI/NRF Research Chair in Inland Fisheries and Freshwater Ecology, South African Institute for Aquatic Biodiversity, Makhanda, South Africa

*Address all correspondence to: alineacosta1@gmail.com

References

[1] De Necker L, Neswiswi T, Greenfield R, van Vuren JHJ, Brendonck L, Wepener V, et al. Long-term water quality patterns of a flow regulated tropical lowland river. Water. 2020;**12**:37

[2] Smit NJ, Vlok W, van Vuren JHJ, du Preez L, van Eeden E, O'Brien GC, et al. Socio-ecological System Management of the Lower Phongolo River and Floodplain Using Relative Risk Methodology. Pretoria (RSA): Water Research Commission; 2016. Report No.: 2185/1/16

[3] Heeg J, Breen CM. Man and the Pongolo Floodplain—A Report of the Committee for Inland Water Ecosystems, National Programme for Environmental Sciences. Pretoria (RSA): Council for Scientific and Industrial Research; 1982. p. 117. Report No.: 58

[4] Dube T, Wepener V, van Vuren JHJ, Smit NJ, Brendonck L. The case for environmental flow determination for the Phongolo River, South Africa. African Journal of Aquatic Science. 2015;**40**(3):269-276

[5] Lankford B, Pringle C, Dickens C, Lewis F, Chhotray V, Mander M, et al. The impacts of ecosystem services and environmental governance on human well-being in the Pongola region, South Africa. London, Norwich, UK and Pietermaritzburg, South Africa: Report to the Natural Environment Research Council. University of East Anglia and Institute of Natural Resources; 2010

[6] Stats SA, Statistics South Africa. 2020. Statistics by Place. Available from: http://www.statssa.gov.za [Accessed 17 July 2020]

[7] Gerber A, Gabriel MJM. Aquatic Invertebrates of South African Rivers: Field Guide. Institute for Water Quality Studies. Department of Water Affairs and Forestry: Pretoria (RSA); 2002. p. 150

[8] Day JA, Harrison AD, De Moor IJ, editors. Guides to the Freshwater Invertebrates of Southern Africa. Vol. 9. Diptera. Pretoria (RSA): Water Research Commission; 2003. p. 211. Report No.: TT 201/02

[9] de Moor IJ, Day JA, de Moor FC, editors. Guides to the Freshwater Invertebrates of Southern Africa. Vol. 7: Insecta I (Ephemeroptera, Odonata and Plecoptera). Pretoria (RSA): Water Research Commission; 2003. p. 300. Report No.: TT 207/03

[10] Griffiths C, Day J, Picker M. Freshwater Life: A Field Guide to the Plants and Animals of Southern Africa. Penguin Random House: South Africa; 2015. p. 1272

[11] Wiggins GB, Mackay RJ, Smith IM. Evolutionary and ecological strategies of animals in annual temporary pools. Archiv für Hydrobiologie. 1980;**58**:97-206

[12] Bird MS, Mlambo MC, Wasserman RJ, Dalu T, Holland AJ, Day JA, et al. Deeper knowledge of shallow waters: Reviewing the invertebrate fauna of Southern African temporary wetlands. Hydrobiologia. 2019;**827**:89-121

[13] Helson JE, Williams DD. Development of a macroinvertebrate multimetric index for the assessment of low-land streams in the neotropics. Ecological Indicators. 2013;**29**:167-178

[14] Feio MJ, Almeida SFP, Craveiro SC, Calado AJ. Diatoms and macroinvertebrates provide consistent and complimentary information on environmental quality. Hydrobiologia. 2007;**169**:247-258

[15] Čiamporová-Zatovičová Z, Hamerlík L, Šporka F, BituŠík P. Littoral benthic macroinvertebrates of alpine lakes (Tatra Mts) along an altitude gradient: A basis for climate change assessment. Hydrobiologia. 2010;**648**:19-34

[16] Thorp JH, Covich AP, editors. Ecology and Classification of North American Freshwater Invertebrates. 3rd ed. London (UK): Academic press; 2010. p. 1021

[17] Villastrigo A, García-Criado F. Response of a littoral macroinvertebrate assemblage to a presistent hydrological drought in a permanent mountain shallow lake from northwestern Spain. Annales de Limnolofie-International Journal of Limnology. 2017;**53**:315-323

[18] Barbour MT, Gerritsen J, Snyder BD, Stribling JB. Rapid Bioassessment Protocols for Use in Streams and Wadeable Rivers: Periphyton, Benthic Macroinvertebrates and Fish. 2nd ed. Washington D.C. (USA): U.S. Environmental Protection Agency, Office of Water; 1999. p. 339. Report No.: EPA 841-B-99-002

[19] Dickens CWS, Graham PM. The South African Scoring System (SASS) version 5 rapid bioassessment method for rivers. African Journal of Aquatic Science. 2002;**27**:1-10

[20] Dube T, De Necker L, van Vuren JHJ, Wepener V, Smit NJ, Brendonck L. Spatial and temporal variation of invertebrate community structure in flood-controlled tropical floodplain wetlands. Journal of Freshwater Ecology. 2017;**32**:1-15

[21] Dube T, de Necker L, Wepener V, Smit NJ, Pinceel T, Mwaijengo GN, et al. A comparison of aquatic macroinvertebrate and large branchiopod community composition between temporary pans of a conservation area and surrounding communal area in South Africa. African Zoology. 2020;**55**:67-77

[22] South African Wetlands Conservation Programme (SA Wetlands Conservation Programme). Ndumo Game Reserve, South Africa, Information Sheet. Pretoria (RSA): Department of Environmental Affairs and Tourism; 1996. p. 33. Report No.: 24/21/3/3/3/15

[23] Scealy JA, Mika SJ, Boulton AJ. Aquatic macroinvertebrate communities on wood in an Australian lowland river: Experimental assessment of the interactions of habitat, substrate complexity and retained organic matter. Marine and Freshwater Research. 2007;**58**:153-165

[24] Dlamini V, Hoko Z, Murwira A, Magagula C. Response of aquatic macro-invertebrates diversity to environmental factors along the lower Komati River in Swaziland. Physics and Chemistry of the Earth. 2010;**35**:665-671

[25] Ferreira M, Wepener V, van Vuren JHJ. Aquatic invertebrate communities of perennial pans in Mpumalanga, South Africa: A diversity and functional approach. African Invertebrates. 2012;**53**:751-768

[26] Senner NR, Moore JN, Seager ST, Dougill S, Kreuz K, Senner SE. A salt lake under stress: Relationships among birds, water levels, and invertebrates at a Great Basin saline lake. Biological Conservation. 2018;**220**:320-329

[27] Al-Shami SA, Rawi CSM, Ahmad AH, Hamid SA, Nor SAM. Influence of agricultural, industrial and anthropogenic stresses on the distribution and diversity of macroinvertebrates in Juru River basin, Penang, Malaysia. Ecotoxicology and Environmental Safety. 2011;**74**(5):1195-1202

[28] Jaganyi J, Salagae M, Matiwane N. Integrating Floodplain Livelihoods into

a Diverse Rural Economy by Enhancing Co-operative Management: A Case Study of the Pongolo Floodplain System, South Africa. Pretoria (RSA): Water Research Commission; 2009. p. 400. Report No.: 1299/1/08

[29] Lankford B, Pringle C, Dickens C, Lewis F, Mander M, Chhotray V, et al. Hydrological modelling of water allocation, ecosystem services and poverty alleviation in the Pongola floodplain, South Africa. Journal of Environmental Planning and Management. 2011;**54**:1237-1260

[30] Food and Agriculture Organization of the United Nations (FAO). The State of World Fisheries and Aquaculture 2018—Meeting the Sustainable Development Goals. Rome, Italy: FAO; 2018

[31] Skelton P. A Complete Guide to the Freshwater Fishes of Southern Africa. Cape Town, South Africa: Struik Publishers; 2001. p. 395

[32] Froese R, Pauly D. FishBase. World Wide Web Electronic Publication. 2020. Available from: www.fishbase.org [Accessed: 01 August 2020]

[33] Merron GS, Bruton MN, la Hausse de Lalouviere P. Implications of water release from the Pongolapoort Dam for the fish and fishery of the Phongolo floodplain, Zululand. Southern African Journal of Aquatic Sciences. 1993;**19**:34-49

[34] Merron GS, Weldrick SR, Kaiser H, Bruton MN. A Fisheries Survey of the Phongolo Floodplain, Zululand, South Africa (December Survey). JLB Smith Institute of Ichthyology Investigational Report No. 45. Grahamstown, South Africa: JLB Smith Institute of Ichthyology; 1994

[35] Merron GS, Weldrick SR, Bruton MN. The Second Fisheries Survey of the Phongolo Floodplain, Zululand, South Africa (February Survey). JLB Smith Institute of Ichthyology Investigational Report No. 46. Grahamstown, South Africa: JLB Smith Institute of Ichthyology; 1994

[36] Merron GS, Weldrick SR, Kaiser H, Bruton MN. The Third Fisheries Survey of the Phongolo Floodplain, Zululand, South Africa (April Survey). JLB Smith Institute of Ichthyology Investigational Report No. 47. Grahamstown, South Africa: JLB Smith Institute of Ichthyology; 1994

[37] Merron GS, Weldrick SR, Kaiser H, Bruton MN. The Fourth Fisheries Survey of the Phongolo Floodplain, Zululand, South Africa (June Survey). JLB Smith Institute of Ichthyology Investigational Report No. 48. Grahamstown, South Africa: JLB Smith Institute of Ichthyology; 1994

[38] Merron GS, Weldrick SR, Kaiser H, Bruton MN. The Fifth Fisheries Survey of the Phongolo Floodplain, Zululand, South Africa (August Survey). JLB Smith Institute of Ichthyology Investigational Report No. 49. Grahamstown, South Africa: JLB Smith Institute of Ichthyology; 1994

[39] Merron GS, Weldrick SR, Kaiser H, Bruton MN. The Sixth Fisheries Survey of the Phongolo Floodplain, Zululand, South Africa (November Survey). JLB Smith Institute of Ichthyology Investigational Report No. 50. Grahamstown, South Africa: JLB Smith Institute of Ichthyology; 1994

[40] Bills R. *Oreochromis mossambicus* (errata version published in 2020). The IUCN Red List of Threatened Species. 2019:e.T63338A174782954. Available from: https://dx.doi.org/10.2305/IUCN.UK.2019-3.RLTS.T63338A174782954 [Accessed: 12 July 2020]

[41] Department of Environmental Affairs and Tourism (DEAT). National Environmental Management:

Biodiversity Act, 2004 (Act No. 10 of 2004). Publication of lists of species that are threatened or protected, activities that are prohibited and exemption from restriction. Government Gazette No. 36375. 16 April 2013. Available from: https://www.environment.gov.za/sites/default/files/gazetted_notices/nemba10of2004_threaten_protected_species_regulations_0.pdf [Accessed: 12 July 2020]

[42] Van der Bank F, Deacon A. Increased backcrossing has reduced the usefulness of morphological and allozyme data for identifying *Oreochromis niloticus*, *O. mossambicus* (Teleostei: Cichlidae) and their hybrids in the Pafuri reach of the Luvuvhu River in the Kruger National Park, South Africa. African Journal of Aquatic Science. 2007;**32**(2):193-196

[43] De Necker L, Manfrin A, Ikenaka Y, Ishizuka M, Brendonck L, van Vuren JHL, et al. Using stable $\delta^{13}C$ and $\delta^{15}N$ isotopes to assess foodweb structures in an African subtropical temporary pool. African Zoology. 2020;**55**(1):79-92

[44] Coetzee HC, Nell W, van Eeden E, de Crom N. Artisanal Fisheries in the Ndumo Area of the Lower Phongolo River Floodplain, South Africa. Koedoe. 2015;**57**(1):1248

[45] McHugh KJ, Smit NJ, van Vuren JHJ, van Dyk JC, Bervoets L, Covaci A, et al. A histology-based fish health assessment of the tigerfish, *Hydrocynus vittatus* from a DDT-affected area. Physics and Chemistry of the Earth. 2011;**36**:895-904

[46] Frost DR. Amphibian species of the world: an online reference. Version 6.0. New York, USA: American Museum of Natural History; 2020

[47] Beebee TJC, Griffiths RA. The amphibian decline crisis: A watershed for conservation biology? Biology Conservation. 2005;**125**:271-285

[48] Stuart SN, Hoffmann M, Chanson JS, Cox NA, Berridge RJ, Ramani P, et al. Threatened Amphibians of the World. Barcelona, Spain: Lynx Edicions; 2008

[49] Frost DR, Grant T, Faivovich J, Bain RH, Haas A, Haddad CFB, et al. The amphibian tree of life. Bulletin of the American Museum of Natural History. 2006;**297**:1-291

[50] Du Preez LH, Carruthers V. Frogs of Southern Africa: A Complete Guide. New York, NY: Penguin Random House; 2007

[51] AmphibiaWeb. AmphibiaWeb. Berkeley, CA, USA: University of California; 2020

[52] Measey GJ. Ensuring a Future for South Africa's Frogs: A Strategy for Conservation Research. Pretoria: South African National Biodiversity Institute; 2011

[53] Minter LR, Burger M, Harrison JA, Braack HH, Bishop PJ, Kloepfer D. Atlas and Red Data Book of the Frogs of South Africa, Lesotho, and Swaziland. Washington, DC: Smithsonian Institution; 2004

[54] Channing A, Rödel M-O. Field Guide to the Frogs & Other Amphibians of Africa. South Africa, Cape Town: Struik Nature, Penguin Random House; 2019

[55] Drinkrow DR, Cherry MI. Anuran distribution, diversity and conservation in South Africa, Lesotho and Swaziland. African Zoology. 1995;**30**:82-90

[56] Minter LR. Frogs and climate change in South Africa. Current Allergy and Clinical Immunology. 2011;**24**:75-78

[57] Graeter GJ, Rothermel BB, Gibbons JW. Habitat selection and movement of pond-breeding amphibians in experimentally fragmented pine forests. Journal of Wildlife and Management. 2008;**72**:473-482

[58] Alexander G, Harrison J, Fairbanks D, Navarro R. Biogeography of the frogs of South Africa, Lesotho and Swaziland. In: Minter LR, Burger M, Harrison JA, Braack HH, Bishop PJ, Kloepfer D, editors. Atlas and Red Data Book of the Frogs of South Africa, Lesotho and Swaziland. Washington, DC: Smithsonian Institution; 2004. pp. 31-47

[59] Phaka FM, Netherlands EC, Kruger DJD, du Preez LH. A Bilingual Field Guide to the Frogs of Zululand. Pretoria: South African Biodiversity Institute; 2017

[60] Minter LR, Netherlands EC, du Preez LH. Uncovering a hidden diversity: Two new species of *Breviceps* (Anura: Brevicipitidae) from northern KwaZulu-Natal, South Africa. Zootaxa. 2017;**4300**(2):195-216

[61] Luque JL, Poulin R. Metazoan parasite species richness in neotropical fishes: Hotspots and the geography of biodiversity. Parasitology. 2007;**134**(6):865-878

[62] Poulin R. Parasite biodiversity revisited: Frontiers and constraints. International Journal of Parasitology. 2014, 2014;**44**:581-589

[63] Adlard RD, Miller TL, Smit NJ. The butterfly effect: Parasite diversity, environment, and emerging disease in aquatic wildlife. Trends in Parasitology. 2015;**31**(4):160-166

[64] Schmidt GD, Roberts LS. Foundations of Parasitology. 8th ed. New York: McGraw-Hill; 2009. p. 701

[65] Alicata JE. Parasites of Man and Animals in Hawaii. Basel: Karger; 1969. pp. 15-26

[66] Wells K, Clark NJ. Host specificity in variable environments. Trends in Parasitology. 2019;**35**(6):452-465

[67] Buchmann K, Bresciani J, Pedersen K, Ariel E, Dalsgaard I, Madsen L. Fish Diseases—An Introduction. Narayana Press: Frederiksberg; 2009. p. 131

[68] Sures B, Nachev M, Selbach C, Marcogliese DJ. Parasite responses to pollution: What we know and where we go in 'environmental parasitology'. Parasites & Vectors. 2017;**10**:65

[69] Erasmus JH, Wepener V, Nachev M, Zimmermann S, Malherbe W, Sures B, et al. The role of fish helminth parasites in monitoring metal pollution in aquatic ecosystems: A case study in the world's most productive platinum mining region. Parasitology Research. 2020;**119**:2783-2798

[70] Acosta AA, Queiroz J, Brandão H, Carvalho ED, Silva RJ. Helminths of *Steindachnerina insculpta* in two distinct stretches of the Taquari River, state of São Paulo, Brazil. Revista Brasileira de Parasitologia Veterinária. 2013;**22**:539-547

[71] Price CE, Peebles HE, Bamford T. The monogenean parasites of African fishes—IV; two new species from South African hosts. Revue de Zoologie et de Botanique Africaines. 1969;**79**:117-124

[72] Scholz T, Vanhove MPM, Smit N, Jayasundera Z, Gelnar M, editors. A Guide to the Parasites of African Freshwater Fishes, Abc Taxa. Brussels: Royal Belgian Institute of Natural Sciences Scientific Publication Unit; 2018. p. 420

[73] Svitin R, Truter M, Kudlai O, Smit NJ, du Preez L. Novel information on the morphology, phylogeny and distribution of camallanid nematodes from marine and freshwater hosts in South Africa, including the description of *Camallanus sodwanaensis* n. sp.

International Journal for Parasitology: Parasites and Wildlife. 2019;**10**:263-273

[74] Hoogendoorn C, Smit NJ, Kudlai O. Resolution of the identity of three species of *Diplostomum* (Digenea: Diplostomidae) parasitising freshwater fishes in South Africa, combining molecular and morphological evidence. International Journal for Parasitology: Parasites and Wildlife. 2020;**11**:50-61

[75] Smit NJ, Joubert A, Lawton SP, Hayes PM, Cook CA. Morphological and molecular characterization of an African freshwater fish trypanosome, including its development in a leech vector. International Journal for Parasitology. 2020;**50**:921-929

[76] Schaeffner BC, van Rooyen D, Gerber R, Scholz T, Smit NJ. Original distribution boundaries expanded: Description of the first native caryophyllidean tapeworm, *Wenyonia gracilis* sp. n. (Cestoda), from southern Africa. Folia Parasitologica. [in press]

[77] Řehulková E, Kičinjaová ML, Mahmoud ZN, Gelnar M, Seifertová M. Species of *Characidotrema* Paperna & Thurston, 1968 (Monogenea: Dactylogyridae) from fishes of the Alestidae (Characiformes) in Africa: New species, host-parasite associations and first insights into the phylogeny of the genus. Parasites & Vectors. 2019;**12**:366

[78] Truter M, Hadfield KA, Weyl OLF, Smit NJ. *Pseudodactylogyrus anguillae* (Yin & Sproston, 1948) from the giant mottled eel *Anguilla marmorata* Quoy & Gaimard, 1824 in the Phongolo River, South Africa: aggressive invader or native gill monogenean? Parasitology Research. [in press]

[79] Truter M, Weyl OLF, Smit NJ. New distribution records for species of *Macrogyrocadtylus* Malmberg, 1957 (Monogenea: Gyrodactylidae) from Southern Africa: supplementary morphological observations and molecular characterisation. International Journal for Parasitology: Parasites and Wildlife. [in press]

[80] Combes C. Parasites, biodiversity and ecosystem stability. Biodiversity and Conservation. 1996;**5**:953-962

[81] Davies AJ, Johnston MRL. The biology of some intraerythrocytic parasites of fishes, amphibia and reptiles. Advances in Parasitology. 2000;**45**:1-107

[82] Netherlands EC, Cook CA, Du Preez LH, Vanhove MPM, Brendonck L, Smit NJ. An overview of the Dactylosomatidae (Apicomplexa: Adeleorina: Dactylosomatidae), with the description of *Dactylosoma kermiti* n. sp. parasitising *Ptychadena anchietae* and *Sclerophrys gutturalis* from South Africa. International Journal for Parasitology: Parasites and Wildlife. 2020;**11**:246-260

[83] Netherlands EC, Cook CA, Du Preez LH, MPM V, Brendonck L, Smit NJ. Monophyly of the species of *Hepatozoon* (Adeleorina: Hepatozoidae) parasitizing (African) anurans, with the description of three new species from hyperoliid frogs in South Africa. Parasitology. 2018;**145**:1039-1050

[84] Netherlands EC, Cook CA, Smit NJ. *Hepatozoon* species (Adeleorina: Hepatozoidae) of African bufonids, with morphological description and molecular diagnosis of *Hepatozoon ixoxo* sp. nov. parasitising three *Amietophrynus* species (Anura: Bufonidae). Parasites & Vectors. 2014;**7**:552

[85] Netherlands EC, Cook CA, Kruger DJD, Du Preez LH, Smit NJ. Biodiversity of frog haemoparasites from sub-tropical northern KwaZulu-Natal, South Africa. International Journal for Parasitology: Parasites and Wildlife. 2015;**4**:135-141

[86] Netherlands EC. Ecology, Systematics and Evolutionary Biology

of Frog Blood Parasites in Northern KwaZulu-Natal. Leuven, Belgium: KU Leuven and North-West University; 2019

[87] Dobson A, Lafferty KD, Kuris AM, Hechinger RF, Jetz W. Homage to Linnaeus: How many parasites? How many hosts? Proceedings of the National Academy of Sciences. 2008;**105**:11482-11489

[88] Netherlands EC. Species Diversity, Habitat Utilization and Blood Parasites of Amphibians in and Around Ndumo Game Reserve, Unit for Environmental Sciences and Management. Potchefstroom: North-West University; 2014

[89] Ellender BR, Weyl OLF. A review of current knowledge, risk and ecological impacts associated with non-native freshwater fish introductions in South Africa. Aquatic Invasions. 2014;**9**:117-132

[90] Wilson JRU, Dormontt EE, Prentis PJ, Lowe AJ, Richardson DM. Something in the way you move: Dispersal pathways affect invasion success. Trends in Ecology & Evolution. 2009;**24**:136-144

[91] Weyl OLF, Ellender B, Wassermann RJ, Truter M, Dalu T, Zengeya TA, et al. Alien freshwater Fauna in South Africa. Chapter 6. In: van Wilgen B, Measey J, Richardson DM, Wilson JR, Zengeya TA, editors. Biological Invasions in South Africa. Springer Series in Invasion Ecology. Vol. 14. Springer International Publishing, New York. 2020. p. 153-183

[92] Dobson M. Replacement of native freshwater snails by the exotic *Physa acuta* (Gastropoda: Physidae) in southern Mozambique; a possible control mechanism for schistosomiasis. Annals of Tropical Medicine and Parasitology. 2004;**98**:543-548

[93] Poulin R, Paterson RA, Townsend CR, Tompkins DM, Kelly DW. Biological invasions and the dynamics of endemic diseases in freshwater ecosystems. Freshwater Biology. 2011;**56**:676-688

[94] Lymbery AJ, Morine M, Kanani HG, Beatty SJ, Morgan DL. Co-invaders: The effects of alien parasites on native hosts. International Journal for Parasitology: Parasites and Wildlife. 2014;**3**:171-177

[95] Wilson JR, Saunders RJ, Hutson KS. Parasites of the invasive tilapia *Oreochromis mossambicus*: Evidence for co-introduction. Aquatic Invasions. 2019;**14**:332-349

[96] Dudgeon D, Arthington AH, Gessner MO, Kawabata Z-I, Knowler DJ, Lêvêque C, et al. Freshwater biodiversity: Importance, threats, status and conservation challenges. Biological Reviews. 2006;**81**:163-182

[97] Moorhouse TP, Macdonald DW. Are invasives worse in freshwater than terrestrial ecosystems? WIREs Water. 2015;**2**:18

[98] Nunes AL, Zengeya TA, Hoffman AC, Measy GJ, Weyl OLF. Distribution and establishment of the alien Australian redclaw crayfish, *Cherax quadricarinatus*, in South Africa and Swaziland. PeerJ. 2017;**5**:e3135

[99] du Preez LH, Smit NJ. Double blow: Alien crayfish infected with invasive temnocephalan in South African waters. South African Journal of Science. 2013;**109**(9/10). Art #2013-0109, 4 p

[100] Smit NJ, Malherbe W, Hadfield KA. Alien freshwater fish parasites from South Africa: Diversity, distribution, status and the way forward. International Journal for Parasitology Parasites and Wildlife. 2017;**6**:386-401

[101] De Necker L. Biodiversity and ecological structures of an African subtropical river and associated

floodplain pans [thesis]. Potchefstroom: North-West University; 2019

[102] Welicky RL, De Swart J, Gerber R, Netherlands E, Smit NJ. Drought associated absence of alien invasive anchorworm, *Lernaea cyprinacea* (Copepoda: Lernaeidae), is related to changes in fish health. International Journal for Parasitology Parasites and Wildlife. 2017;**6**:430-438

[103] Appleton CC, Miranda NAF. Two Asian freshwater snails newly introduced into South Africa and an analysis of alien species reported to date. African Invertebrates. 2015;**56**:1-17

[104] CABI, Invasive Species Compendium. *Cherax quadricarinatus* (redclaw crayfish). 2011. Available from: https://www.cabi.org/isc/datasheet/89135#todescription [Accessed: 4 August 2020]

[105] Tavakol S, Luus-Powell WJ, Smit WJ, Baker C, Hoffman A, Halajian A. First introduction of two Australian temnocephalan species into Africa with an alien host: Double trouble. Journal of Parasitology. 2016;**102**(6):653-658

[106] Brown C, Joubert A, Tlou T, Birkhead A, Marneweck G, Paxton B, et al. The Pongola floodplain, South Africa—Part 2: Holistic environmental flows assessment. Water SA. 2018;**44**(4):748-759

[107] Hitayese P, Wale E, Ortmann GF. Assessing agricultural land-use change in the midlands region of KwaZulu-Natal, South Africa: Application of mixed multinomial logit. Environment, Development and Sustainability. 2016;**18**:985-1003

[108] Smit HE, Bouwman H, Le Sueur D. DDT and community health. Journal of Comprehensive Health. 1991;**3**(1):175-178

[109] Sereda BL, Meinhardt HR. Insecticides Contamination of the Water Environment in Malaria Endemic Areas of KwaZulu-Natal (SA): The Risk of Insecticide Resistance Development in Malaria Vectors. WRC Report No. 1119/1/03. Pretoria: Water Research Commission; 2003

[110] Bouwman H, Coetzee A, Schutte CHJ. Environmental and health implications of DDT-contaminated fish from the Pongolo flood plain. Journal of African Zoology. 1990;**104**:275-286

[111] Bouwman H, Cooppan RM, Becker RT, Ngxongo S. Malaria control and levels of DDT in serum of two populations in KwaZulu-Natal. Journal of Toxicology and Environmental Health. 1991;**33**(3):141-155

[112] Gerber R, Wepener V, Smit NJ. Application of multivariate statistics and toxicity indices to evaluate the water quality suitability for fish of three rivers in the Kruger National Park, South Africa. African Journal of Aquatic Science. 2015;**3**:247-259

[113] Gerber R, Smit NJ, van Vuren JHJ, Nakayama SMM, Yohannes YB, Ikenaka Y, et al. Bioaccumulation and human health risk assessment of DDT and other organochlorine pesticides in an apex aquatic predator from a premier conservation area. Science of the Total Environment. 2016;**550**:522-533

[114] Volschenk CM, Gerber R, Mkhonto MT, Ikenaka Y, Yohannes YB, Nakayama S, et al. Bioaccumulation of persistent organic pollutants and their trophic transfer through the food web: Human health risks to the rural communities reliant on fish from South Africa's largest floodplain. Science of the Total Environment. 2019;**685**:1116-1126

[115] King N, Strydom H, Retief F. Environmental Management in South Africa. 3rd ed. Cape Town: Juta Publishing; 2018

[116] Patterson A. Biological diversity. In: King N, Strydom H, Retief F, editors. Environmental Management in South Africa, 3rd ed. Cape Town: Juta Publishing; 2018

[117] Goosen M, Blackmore AC. Hitchhikers' guide to the legal context of protected area management plans in South Africa. Bothalia. 2019;**49**(1):a2399

[118] Alberts R, Retief F, Roos C, Cilliers D, Arakele M. Re-thinking the fundamentals of EIA through the identification of key assumptions for evaluation. Impact Assessment and Project Appraisal. 2020;**38**(3):205-213

[119] Roos C, Cilliers D, Retief F, Alberts R, Bond A. 'Regulators' perceptions of environmental impact assessment (EIA) benefits in a sustainable development context. Environmental Impact Assessment Review. 2020;**81**:106360

[120] Retief F, Morrison-Saunders A, Geneletti D, Pope J. Exploring the psychology of trade-off decision making in EIA. Impact Assessment and Project Appraisal. 2013;**31**(1):13-23

[121] Retief FP, Alberts RC, Roos C, Cilliers DC, Siebert F. Identifying key risks to the effectiveness of privately protected areas (PPAs) through the application of theory of change (ToC). Biological Conservation. 2020 [under review]

[122] Musavengane R, Simatele DM. Community-based natural resource management: The role of social capital in collaborative environmental management of tribal resources in Kwazulu-Natal, South Africa. Development Southern Africa. 2016;**33**(6):806-821

[123] Nkambule SS, Buthelezi HZ, Munien S. Opportunities and constraints for community based conservation: The case of Kwazulu-Natal Sandstone Sourveld grassland, South Africa. Bothalia—African Biodiversity and Conservation. 2016;**46**:1-8

[124] Pollard S, Cousins T. Community-Based Governance of Freshwater Resources in Southern Africa. 2018. Water Research Commission Report, TT328/08

Chapter 3

The Predicament of Macaque Conservation in Malaysia

Siew Shean Choong, Mimi Armiladiana Mohamad, Li Peng Tan and Ruhil Hayati Hamdan

Abstract

Macaques are commonly found in Malaysia, with the current existing three species placed between endangered to least concern status under the IUCN Red List, namely the stump-tailed macaque (*Macaca arctoides*), pig-tailed macaque (*Macaca nemestrina*), and the notorious long-tailed macaque (*Macaca fascicularis*). The species classified under the endangered and vulnerable group are facing threats mainly from the loss of habitat. Conversely, species that are categorized as least concerned are often cited at the top of human-wildlife conflicts reports in various countries, although they too are facing pressure from habitat loss. There are different methods employed to control the fast-growing population of these species, calling for different levels of investment in terms of resources. It is of great interest to understand the disparities between these species, as they are able to adapt to environmental changes and some find ways to survive in alternative localities, including urban areas. The proximity of macaques to human dwellings raises a public health concern through the transmission of zoonotic diseases. More scientific studies are imperative in order to further understand the needs of these animals for continued survival and co-existence with humans and other animals in the ecosystem. Urgent efforts must be taken to preserve the macaque's natural habitats while creating the public awareness on the predicament of these species. The focus should be on human-wildlife conflicts todispute the existing false impression that all macaques are on equal ground and abundance in numbers.

Keywords: macaque conservation, non-human primate conservation, public health, zoonoses

1. Introduction

Malaysia is located in the equatorial region where most parts of the natural landscape are covered by the tropical rainforests. The country is well known for its rich flora and fauna biodiversity. Non-human primate species found natively in Malaysia include the great ape: Bornean orangutan (*Pongo pygmaeus*); lesser ape: Agile gibbon (*Hylobates agilis*), White-handed gibbon (*Hylobates lar*), Bornean gibbon (*Hylobates muelleri*), Siamang (*Symphalangussyndactylus*); old world monkeys: Banded leaf monkey (*Presbytis femoralis*), White-fronted langur (*Presbytis frontata*), Gray leaf monkey (*Presbytis hosei*), Red leaf monkey (*Presbytis rubicunda*), Silver leaf monkey (*Trachypithecuscristatus*), Dusky leaf monkey (*Hylobates funereus obscurus*), Long-tailed macaque (*Macaca fascicularis*), pig-tailed macaque (*Macaca nemestrina*), stump-tailed macaque (*Macaca arctoides*); the lorisids: Sunda slow loris (*Nycticebuscoucang*), Kayan slow loris (*Nycticebuskayan*); lastly the Western tarsier (*Tarsiusbancanus*).

The focus of this chapter is management aspects of macaques as the conservation of these species is generally neglected in the country. They were at the extreme end of the conservation status as indicated by the IUCN Red List of Threatened Species, where long-tailed macaques are categorized as least concerned, while the pig-tailed macaques are endangered. Furthermore, reports, conservation efforts, and research centers on macaques are almost unheard of, except human-macaque conflicts that mainly involved the long-tailed macaques. It may be fortunate that young macaques do not appear cute and cuddly compared to that of orangutans and leaf monkeys, they are less often reported to be illegally traded or poached. However, this may also be one of the reasons they are not seen as "attractive" subjects of for research and conservation. Therefore, this chapter aspires to highlight the plight of macaques.

2. The macaques population in Malaysia

Although the long-tailed macaques are considered as pests in Malaysia, the species is listed as vulnerable under the IUCN Red List [1]. This is particularly true as their natural habitats are diminishing. According a report by Wicke et al. [2], Malaysia has lost 20 percent of forest land within a 30-year span due to anthropogenic activities. The downward trend of primary forest cover continues. The 91.9% of the cover remained in 2011 was further reduced to 83% in 2020 [3]. Although deforestation was significantly controlled over the past decade, the conversion of forest land is expected to continue creating more fragmented forest areas and forest edges [4], which are suitable habitats for long-tailed macaques. Subsequently, this would further escalate the frequencies of human-wildlife conflicts. Continuous mass removal of the long-tailed macaque which has been adopted to minimize human-wildlife conflicts subject the species to a risk of losing genetic diversity [5]. In fact, hybridization of populations and potential inbreeding depression of the populations could potentially occur if translocation operations intensified [6]. Additionally, long-tailed macaques in Peninsular Malaysia are morphologically assigned to two subspecies, namely *M. f. fascicularis* and *M. f. argentimembris* [7].

On the other hand, pig-tailed macaques and long-tailed macaques are generally regarded as crop raiders and, therefore, have more direct negative interactions with people. However, pig-tailed macaques have proved to be beneficial to mankind. They have traditionally been kept and trained to harvest fruits, especially coconuts, and forest products for over a century in Malaysia and other countries in Southeast Asia as the species is bigger and has more physical strength than long-tailed macaques. A recent research has indicated that the presence of pig-tailed macaques benefits the oil palm plantations by acting as a biological control for rodents that cause the industry monetary losses of US$930mil (RM3.9bil) every year [8]. In comparison, the damage caused by the pig-tailed macaques on the oil palm crops is relatively minimal.

Pig-tailed macaques are listed as endangered in Malaysia since 2009, although it is still categorized as vulnerable when the entire population worldwide is considered [9]. It is of interest to note that both long-tailed macaques and pig-tailed macaques shared similar natural habitats and are omnivorous. However, the latter is more sociable towards humans, less aggressive, and more habituated. Currently, efforts to conserve the species in the wild are not viewed as critical. The breeding of pig-tailed macaques is generally undertaken as the animals are needed for fruit plucking purposes. Yet, this effort is not nationwide, but concentrated only in the East Coast of Peninsular Malaysia and mostly unable to sustain the population if the animals go extinct in the wild.

Stump-tailed macaques are found only in the North-western region in Peninsular Malaysia, specifically at the Wan Kelian forest areas in Perlis State Park. In fact, the global geographical range published by IUCN indicated that this is the most southern region where stump-tailed macaques can be detected. Their presence could be found further north in Thailand, Myanmar, Laos, Cambodia, Vietnam, Bangladesh, and China. Distributions of the stump-tailed macaques are similarly in habitat pockets as in Malaysia, not widespread throughout the countries listed, and the populations are mostly declining [10]. Unfortunately, wildlife censuses have been excluding macaques as the species of interest in research, thus making more recent data unavailable. Despite its vulnerable status under the IUCN Red List [11], these macaques received the least attention from the government, public, and even researchers compared to the other two macaque species in Malaysia [7, 12]. Small population size, movement between country borders (Malaysia and Thailand) which that is prone to heavy poaching, and their tendencies to avoid humans [12] further complicate research of this species. It is interesting to note that all three macaque species: stump-tailed macaques,pig-tailed macaques and long-tailed macaques are found sympatrically within the Perlis State Park Forest Areas [12, 13]. However, interspecies associations were not recorded between the three species [12, 14].

3. Macaques involvement in human-wildlife conflicts

Among the non-human primate species found in the country, the long-tailed macaques and pig-tailed macaques can be found in areas that often overlap with high anthropogenic activity areas such as plantations, the secondary forest surrounding human settlements, besides their natural habitats in the wild [1, 9]. Between the two species, the long-tailed macaques are most sighted and involved in human-wildlife conflicts in both rural and urban areas, accounting for between 35 and over 65 percent of conflict reports received by the wildlife department [15, 16]. This species has successfully adapted to the human settlements and continues to multiply at alarming rates, which further contributes to human-wildlife conflicts. The damages caused by macaque related conflicts include injuries such as scratches and bites sustained by people during the encounter with macaques, destruction of properties and materials within when macaques enter and ransack the properties, disturbance to residents and tourists due to the animals' aggressive behavior [17] to snatch and steal when needed.

In general, the main reason for human-wildlife conflicts caused by long-tailed macaques is food motivation. Conversely, by and large, these past behaviors were triggered by human actions: feeding of animals by "good samaritans", an improper garbage disposal that allow animals to forage for scraps, destruction of natural habitats due to deforestation and agricultural activities, and encroachment of human settlement into the forested areas [18]. The authorities manage the macaque human-wildlife conflict through several approaches including culling, public education, and awareness creation program (particularly with the help of non-government organizations), and translocation of the animals (Pers. Observation).

A study on the dietary composition of wild stump-tailed macaques indicated that they mainly consume plant materials from the forest [19]. Unlike its cousins, the long-tailed macaques and pig-tailed macaques are more often spotted at human concentrated areas, and at times orchards, plantations, and even garbage collection sites. Essentially, absence of human-wildlife conflict reports on stump-tailed macaques explains why public knowledge on the species is minimal.

4. Zoonotic diseases in macaques

The genetic relatedness of non-human-primates to humans generally gives the perception that interaction with non-human-primates poses higher zoonotic risks compared to other animal species. The natural habitats of non-human-primates are located at the equatorial zone worldwide, expanding across tropical rainforest in Southeast Asia, West and Central Africa, and South and Central America, also smaller patches in areas adjacent to these. Additionally, non-human-primates can be found in captivity such as in zoological gardens, rescue centers, and animal research facilities. Therefore, it is logical to acknowledge that a large portion of the human population has direct or indirect contact with non-human primates, thus making zoonotic disease spillover a major public health concern. Additionally, interactions of non-human primates with livestock have been reported to initiate multidirectional pathogen transmission between these species, and ultimately lead to spillover to the human population [20]. This is further evident from host-pathogen databases analysis among primates showed that sympatric host species have high probabilities to share parasite species [21].

Interestingly, consumption of bushmeat from non-human primates is not widely practiced or acceptable by the local communities, mainly due to religious practices among the majority of the population. This might have significantly reduced the probability of zoonotic diseases transmission and potential mutagenic changes in the pathogens, particularly viruses. Most popular bushmeat in Malaysia is reportedly from wild pigs (*Sus scrofa*), bearded pigs (*Sus barbatus*), deer (including Sambar deer (*Cervus unicolor*), barking deer (*Muntiacus muntjac*) and mousedeer (*Tragulus kanchil* and *Tragulus napu*), Malayan porcupine (*Hystrix brachyura*), bats, and others [22–25]. On the other hand, sun bears (*Helarctos malayanus*)were hunted for their bile to cater for traditional medicines [26]. Some examples of zoonotic or potential zoonotic diseases reported in macaques are discussed hereunder. However, a comprehensive description of zoonotic disease risks from macaques is still lacking. The degree of impact of macaques on possible future epidemics needs to be elucidated with further studies utilizing the one health approach integrating data on human, animal, and environmental health. Epidemiology of diseases often stimulates research in wildlife species, especially in the recent years with emerging zoonotic diseases suspected to have originated from wildlife as evidenced from a large number of scientific articles published [27–30]. Conversely, conservation of these species could result in the preservation of ecosystem integrity and creating a buffer zone against novel disease outbreaks [31]. The outcomes from studies of these species may help the conservation authorities to strategize the long-term plans and inform the national policies on effective management and conservation of macaques species in Malaysia.

5. Bacterial diseases

Fecal samples from wild long-tailed macaques and pig-tailed macaques involved in human-wildlife conflicts areas in the Lopburi district of Thailand were found to carry *Esherichai coli*, *Staphylococcus spp*, and *Salmonella* spp. [32]. In another study carried out in the Wulongkou Scenic Area, Henan Province of China discovered that about a quarter of over 400 fecal samples contained *Shigella spp*., *Escherichia coli*, *Klebsiella pneumoniae*, and *Leptospira spp*. Other bacteria detected in lesser prevalence included *Campylobacter jejuni*, *Salmonella spp*., *Staphylococcus aureus*, *Streptococcus pneumonia*, *Yersinia spp*., and *Hafnia paralvei*. Among these bacteria, *Salmonella*, *Shigella*, *E. coli*, *C. jejuni*, and *Yersinia* are zoonotic i.e. shared with humans and other animals.

In a preliminary study of zoonotic pathogens in captive pig-tailed macaques detected *Neisseria spp.* in all four swab samples, namely nasal, buccal, throat, and anal, from 30 individuals. Other bacteria isolated were *Pasturella spp.* and *Moraxella spp.* from the nasal swabs, and *Stenotrophomonas* sp. or *Acinetobacter* sp. from the buccal swabs. Whereas for the anal swabs, *Pasturella spp.*, and *Streptobacillus spp.* were detected, besides *Neisseria spp.* [33]. The different species isolated in the previous studies with the current research may be due to the fact that the former was from wild animals, and fresh droppings were collected, while the latter was from captive macaques and collected directly from multiple orifices of the animals.

6. Viral diseases

RNA viruses are considered a major threat among emerging infectious diseases at the human-non-human primate interface [34], and studies conducted in Malaysia are very much concentrated on this group of viruses. As discussed before, long-tailed macaques are involved in most of the human-macaque conflicts and often these issues were dealt with through capture and relocation to deep forest areas [35, 36]. Therefore, researchers took the opportunity to investigate the viruses carried by these animals to determine if these animals could serve as a reservoir host for these pathogens and pose significant disease threats to human and other animal species in the habitat.

In a study by Ain-Najwa et al. [37] using archived long-tailed macaque samples, where sera were used to detect West Nile virus (WNV) antibody through competitive enzyme-linked immunosorbent assay (c-ELISA), and WNV RNA from oropharyngeal swabs via RT-PCR. Results showed that the macaques were all negative for WNV RNA, yet high WNV antibody prevalence was observed. These results may indicate that the macaques are exposed to WNV from other animals in their habitat, yet infectivity was low, and they may not serve as a reservoir to WNV in the wild.

In Malaysia, Zika virus (ZIKV) was detected in 0.2% of patients with clinical signs corresponding to Zika virus infection during the Malaysia ZIKV surveillance between June 2015 and December 2017 after the declaration of the Public Health Emergency of International Concern (PHEIC) by World Health Organization. The source of infection was undetermined and possible zoonotic transmission from wildlife species, such as macaques, was suspected [38]. A total of 234 long-tailed macaques trapped from multiple sites throughout Peninsular Malaysia in the Wildlife Disease Surveillance Program were evaluated for ZIKV prevalence. The researchers were unable to detect ZIKV RNA from any of the macaques sampled, and only 1.3% showed seropositive for neutralizing antibodies. Thus, the study concluded that long-tailed macaques are not likely to be reservoirs for the Zika virus in Malaysia [39].

On the other hand, Malaysia has experienced massive outbreaks of Chikungunya virus (CHIKV) infection in humans in Malaysia between 1998 and 1999 [40] and cases have been reported since with sporadic surges of infections recorded [41]. Researches were conducted to elucidate the potential of macaques in maintaining the Chikungunya virus during inter-epidemic periods, to explain the sporadic disease occurrences, as the virus has been isolated from monkeys in Africa [42]. A study carried out by Sam et al. (2015) found that viraemia among the wild long-tailed macaques tested was not only lacking but also the seroprevalence rate was low. Therefore, it was concluded that long-tailed macaques living at the human-wildlife conflict areas would have played a minor role in CHIKV transmission, if any, during CHIKV outbreak episodes [36]. In fact, later work by Suhana et al. [43] suggested that CHIKV detected in long-tailed macaques may be a spillover of the virus from humans, based on molecular characterization and phylogenetic analysis of the isolates.

Macacine herpesvirus 1 (MaHV1), commonly known as B virus, has been detected among wild macaques in Asia [44]. However, there is minimal information regarding MaHV1 in macaques of Malaysia. To date, there was only one report indicating that 39% of wild long-tailed macaques sampled from six different states in the country during wildlife management program under the Department of Wildlife and National Parks, Peninsular Malaysia was shedding MaHV1 DNA [45]. While animals from different age groups were detected to shed the virus, through PCR of urogenital and oropharyngeal swabs, seroprevalence through ELISA was highest among the adults [45]. The seroprevalence result corresponded with a previous study from Bali, Indonesia, where most adults were expected to have been infected or exposed to MaHV1 [46]. MaHV1 is designated as Biosafety Level 4 (BSL-4) pathogen because humans with untreated MaHV1 infection have over 70% mortality rate [47]. With the high human-macaque conflict reported in Malaysia, wildlife officers and rangers carried out many translocation operations as a mitigation approach; also reports of bites and scratches from macaques are rather common among people that reside in or visited areas with macaques. In fact, some persons bitten and scratched by macaques may have experienced MAHV1 infection yet did not experience clinical signs of infections [48]. However, peculiar reports of human MaHV1 infections were confined to personnel working with macaques or macaque tissues in a husbandry or research environment in the US and Europe. This could be explained by the fact that wild macaques are not shedding the virus in such high concentrations compared to laboratory animals, as the latter may be constantly confined in captivity and exposed to high-stress conditions, where they are handled and manipulated. While their wild cousins are generally free in their natural habitat, and people only have random and occasional encounters with these animals [49]. Even so, workers in contact with wild macaques are recommended to put on appropriate personal protective equipment, as capture and translocation efforts may induce stress on these animals and increased viral shedding or reactivation of infection that could potentially infect the workers [45].

In a preliminary study [33] of captive pig-tailed macaques showed the prevalence of several RNA viruses, such as retrovirus, influenza virus, and lyssavirus, through RT-PCR of the buffy coat. Retrovirus was detected in all individuals sampled, followed by influenza virus (56.6%), then lyssavirus at 13.3%. Further investigation using cell culture and nested PCR to detect Simian Foamy virus (SFV) from these samples as SFV has been reported to infect nearly all captive and free-ranging macaques in Asia [50, 51]. HeLa cell culture demonstrated cytopathogenic effects (CPEs) such as being refractile, detached from the culture surface, floating, clumping, increased in cell distance after just the first passage, and foamy appearance was observed in the cells under high magnification by the third passage [33]. However, detection of SFV using nested PCR targeting the Pol genes and LTR genes from extracted trypsinized tissue culture samples did not turn up positive results. Therefore, further work is required to determine if CPE resulted from SFV or other viruses.

Overall, evidence that wild and captive macaques in Malaysia serve as a reservoir for zoonotic viruses is still lacking. Zoonotic infections from macaque to human is very much understudied, and often prevalence research on pathogens in macaques are not incorporated with sampling from the keepers, owners, and other personnel in contact, such as the wildlife rangers. Additionally, pig-tailed macaque owners surveyed did not report illness related to animals kept, as these animals often share the living quarters, food, and drinks with their owners. However, it is safe to say that personnel in contact with macaques should take the necessary precautions to minimize infection transmission from these animals.

7. Parasitic diseases

The most prominent zoonotic parasite reported in macaques is *Plasmodium knowlesi*, recognized as the fifth cause of human malaria, which is transmitted from animals to humans through Anopheles mosquito vector. Knowlesi malaria has now topped the number of human malaria cases reported across most states in Malaysia, especially in Sabah and Sarawak [52, 53]. This situation may be as a result of increased encroachment of human settlement into the forested areas [54], and advancement of malaria diagnosis to molecular method instead of the conventional microscopy detection of the stained blood smear [55].

On another note, gastrointestinal (GI) parasites from macaques are often neglected although these organisms may cause detrimental consequences in humans. A report from Baluran National Park at East Java, Indonesia indicated 89% of fecal samples collected from wild long-tailed macaque were positive of GI parasite, and protozoal infection was slightly higher (89%) compared to helminth (83%). The study found that the most prevalent GI parasite in the macaques is *Trichostrongylus*sp (66%), the next highest parasite is *Entamoeba* sp. (53%), and followed by *Strongyloides* sp. (32%), *Blastocystis* sp. (32%), *Trichuris* sp. (17%), *Giardia* sp. (10%) and *Enterobius* sp. (3%) [56]. Conversely, the prevalence of GI parasites infection was lower at the Kosumpee Forest Park, MahaSarakham, Thailand, where only 35.11% of the fecal samples were positive, including *Strongyloides* spp. (15.27%), *Trichuris* spp. (22.9%), hookworm (4.58%) and *Ascaris* spp. (1.53%) [57].

A comprehensive project was undertaken to investigate GI parasites in Malaysia's non-human primates from the wild, and animals living in urban habitats, and the ones in captivity. This study examined a total of 12 local non-human primate species and illustrated at least 44 species of GI parasites were detected, including seven species of protozoans, 26 species of nematodes, five species of cestodes, five species of trematodes, and one species of pentastomida. The GI parasite distributions were not significantly different between the three groups, and the most prevalent GI parasite was Ascaris spp. (49.7%), followed by *Oesophagostomum* spp. (26.9%) [58]. A study specifically looking into captive pig-tailed macaques showed that an overall GI parasite prevalence rate of 52%. Among the species, five species belonged to Nematoda viz. *Anatrichosoma* sp., Capillaridae, *Strongyloides* sp., *Trichostrongylus* sp., and *Trichuris* sp. Only one Trematoda species was detected, which is *Paramphistomum* sp. The most common GI parasites are *Trichuris* sp. (38%), followed by *Trichostrongylus* sp. (24%), *Paramphistomum* sp. (14%), *Anatrichosoma* sp. and *Strongyloides* sp. (10%) each, and lastly Capillaridae (5%). It should be noted that about one-third of animals tested had double GI parasite infection (33%), 14% of the infection was single, and 5% had a triple infection. Most of the macaque owners did not administer anthelmintics to their animals as preventive medicine. The authors also examined thick blood smears from these captive pig-tailed macaques and found one sample positive for filaria nematode [59].

On the whole, the GI parasites and haemoparasite identified in these studies are of known public health importance and zoonotic concern that needs to be seriously addressed, specifically raising awareness of people in close contact with macaques.

8. Population control

Due to the success of the long-tailed macaques in adapting to human settlements, particularly in the urban areas, they are the culprit in most reported human-wildlife conflict cases in Malaysia, up to 65% of total annual case reports, compared to any other wildlife species [15–17]. It is most often created serious public nuisance and concerns on the animals causing property damage and bodily harm to people encountered during the conflict episodes. The short-term solution most often resorted to is population control, in the hope to reduce the occurrence of conflicts.

Currently, an effective contraceptive method, besides capturing and physical handling of the animals for surgical and non-surgical neutering methods, is lacking. Zona pellucida vaccination, oral contraceptives are temporary and require reapplication, which is troublesome, labor intensive, and recurring cost. Neutering needs animals to be captured, especially for the females as surgical methods involve laparotomy. The effort starts with procuring and setting up suitable traps, then to restrain, anesthetize, and application of chemical or surgical methods for permanent sterilization.

The most frequently used method for sterilization of the male macaques is castration that involves the removal of the testicles, and this procedure does not require an invasive procedure into the abdominal cavity. Compared with the procedures in the females, often healing time is much quicker and may not require an extended holding period. In a report by Karuppannan et al. [60], non-surgical castration through intraepididymal injections of ethanol-formalin mixture to induce tubular blockage resulted in over 90 percent (32/35) success among the animals tested. This method is labor intensive, requires the animals to be caught first, training of staff and precision during the injection process. Furthermore, this method is most suitable for adult males as epididymis in juveniles and subadults are small and difficult to locate to ensure accurate injection of the ethanol-formalin mixture. This chemical castration method is compared to surgical castration that requires surgical skills and can only be performed by veterinarians, yet the age of animals is usually not an issue for successful removal of testis. However, the chemical approach can be done within a shorter period and the males are expected to sustain their sexual behavior as the testicular tissues remained intact [60].

The impact of castration on male macaques is still debatable. Castration does not appear to impact the social interactions between male Japanese macaques (*Macaca fuscata*) in the group. Instead of linear hierarchy as in the intact males, castrated males are less aggressive and have a more lateral relationship with one another [61]. Thus, Takeshita et al. [61] recommend that castration can be adopted as an effective population control measure. On the other hand, studies indicated possible dental health issues where castrated rhesus macaques (*Macaca mulatta*) that lived till old age have greatly receded alveolar bone with signs of periodontitis more severe than in intact old males, as well as severe temporomandibular joint osteoarthritis in the former [62]. The Department of Wildlife and National Parks would adopt the chemical castration described above as this method would not affect male hormone levels in the animals [60].

On the other hand, sterilization in females will definitely necessitate penetration of the abdominal cavity for removal of the ovaries, and/or the uterus. The length of incision and operation period depends on the method chosen, through laparotomy or laparoscopic approach. Surgical methods have been reported in other countries, the caveat of laparotomy for ovariohysterectomy in the female would require the animals to be kept for at least three to 4 days before release to ensure the suture site has healed. This will require facilities to temporarily house the animals. The use of laparoscopy may alleviate this problem, where tubectomy, removal of the Fallopian tube, ovariectomy (removal of ovaries) in females; and vasectomy in males can be conducted. If done correctly, only two to three small and bloodless (or minimal bleeding) incisions are required to access the reproductive organs [63]. The use of a laparoscope has minimized the length of the abdominal incision, especially in females [64]. Nevertheless, the downside of laparoscopic procedures is costly equipment, the requirement of trained staff, and electricity supply.

9. Conservation of macaques in Malaysia

Malaysia is inhabited by ≥25 non-human primate species from five families, one of the most diverse primate faunas on earth. Unfortunately, most of these primates are threatened with extinction due to habitat loss, degradation and fragmentation, hunting, and the synergies among these processes. Despite the charisma and cultural importance of primates, the significance of primates in ecological processes such as seed dispersal, and the robust development of biodiversity-related sciences in Malaysia, there is relatively little research specifically focusing on wild primates since the 1980s. Forest clearing for plantation agriculture has been a primary driver of forest loss and fragmentation in Malaysia. Selective logging has also negatively impacted the primates. However, these impacts vary across primate taxa. Previously-logged forests were important habitats for many Malaysian primates. Malaysia is crossed by a dense road network, which fragments primate habitats, facilitates further human encroachment into forested areas, and causes substantial mortality due to road kills.

Primates in Malaysia are hunted for food or subjected to retaliatory or pre-emptive killing as pests, trapped for translocation to minimize human-wildlife conflicts, and captured for illegal trade as pets. Additionally, translocation operations should consider conservation of the unique evolutionary lineages of the macaque species, particularly the long-tailed macaques found to be of two distinctive subspecies [5]. Further research on the distribution, abundance, ecology, and behavioral biology of Malaysian primates is needed to inform effective management interventions. Outreach and education are also essential to reduce primate-human conflicts and illegal trade targeting primates as pets. Ultimately, researchers, civil organizations, government authorities, and local and indigenous communities in Malaysia must work together to develop, promote and implement effective strategies to protect Malaysian primates and their habitats.

Of the three macaque species, long-tailed macaques seemed to be able to adapt well within human settlements, despite a high number of human-wildlife conflicts reported [15]. The major conservation challenge facing macaque is habitat loss, degradation, and fragmentation resulting from forest clearing for plantation agriculture, selective logging, and a dense network of roads connecting many cities and townships in the country [65]. Macaques may not be popular as bushmeat, but are also trapped or hunted for illegal trade as pets [65]. Further research on habitat needs for all macaque species is imperative in order to understand the disparity of population density between the species, despite the similarities of natural habitats, diet, and behavior. A good example is a study conducted by Holzner et al. [66] citing the significant changes in sociality behavior of pig-tailed macaques that visit oil palm plantations in Malaysia, which may debilitate individual fitness and infant survival. This proves that despite the ability of pig-tailed macaques to temporarily adapt to human-altered habitats, the proximity of forest is vital for the survival of the species. Research done in recent years indicated an urgent need for macaque conservation strategies to preserve the remaining and segregated pig-tailed macaque and stump-tailed macaque populations involving the authorities, local communities, and general public [12, 67]. The authorities and non-government organizations are urged to increase public awareness on macaque species, particularly their roles in the ecosystem, as little is known about the species.

Author details

Siew Shean Choong*, Mimi Armiladiana Mohamad, Li Peng Tan
and Ruhil Hayati Hamdan
Faculty of Veterinary Medicine, University Malaysia, Kota Bharu, Kelantan,
Malaysia

*Address all correspondence to: shean.cs@umk.edu.my

References

[1] Eudey, A., Kumar, A., Singh, M. &Boonratana, R. (2020) *Macaca fascicularis*. The IUCN Red List of Threatened Species 2020: e.T12551A 17949449. [Internet]. Available from: https://dx.doi.org/10.2305/IUCN.UK.2020-2.RLTS.T12551A17949449.en [Accessed: 2020-12-16]

[2] Wicke B, Sikkema R, Dornburg V, Faaij A. Exploring land use changes and the role of palm oil production in Indonesia and Malaysia. Land Use Policy. 2011;**28**(1):193-206

[3] Global Forest Watch (2021) Primary forest loss in Malaysia. https://gfw.global/3gtXC2v

[4] Omran A, Schwarz-Herion O. Deforestation in Malaysia: The Current Practice and the Way Forward. In: Omran A, Schwarz-Herion O, editors. Sustaining our Environment for Better Future. Singapore: Springer; 2020. DOI: 10.1007/978-981-13-7158-5_11

[5] Abdul-Latiff MAB, Ruslin F, Faiq H, Hairul MS, Rovie-Ryan JJ, Abdul-Patah P, et al. Continental monophyly and molecular divergence of peninsular Malaysia's. Macaca Fascicularis fascicularis. 2014;**2014**:897682

[6] DeSalle R, Amato G. The expansion of conservation genetics. Nature Reviews. Genetics. 2004;**5**(9):702-712. DOI: 10.1038/nrg1425 PMID: 15372093

[7] Abdul-Latiff MAB, Abdul-Patah P, Yaakop S, Md-Zain BM. Aiding pest control management of long-tailed macaques (*Macaca fascicularis fascicularis*) in Malaysia by using molecular markers of mitochondrial DNA. AIP Conference Proceedings. 2017;**1891**:020003. DOI: 10.1063/1.5005336

[8] Holzner A, Ruppert N, Swat F, Schmidt M, Weiß BM, Villa G, et al. Macaques can contribute to greener practices in oil palm plantations when used as biological pest control. Current Biology. 2019;**29**(20):R1066-R1067

[9] Ang, A., Boonratana, R., Choudhury, A. &Supriatna, J. (2020) Macaca nemestrina. The IUCN Red List of Threatened Species 2020: e.T12555A 181324867. [Internet]. Available from: https://dx.doi.org/10.2305/IUCN.UK.2020-3.RLTS.T12555A181324867.en. [Accessed: 2020-12-16]

[10] Molur S, Brandon-Jones D, Dittus W, Eudey A, Kumar A, Singh M, et al. The Status of South Asian Primates: Conservation Assessment and Management Plan (CAMP) Workshop Report. Coimbatore, India: Zoo Outreach Organisation/CBSG-South Asia; 2003

[11] Chetry, D., Boonratana, R., Das, J., Long, Y., Htun, S. & Timmins, R.J. 2020. Macaca arctoides. The IUCN Red List of Threatened Species. Gland, Switerland: IUCN; 2020: e.T12548A185202632. https://dx.doi.org/10.2305/IUCN.UK.2020-3.RLTS.T12548A185202632.en. Downloaded on 28 June 2021

[12] Syamil AR, Mohd-Ridwan AR, Amsah MA, Abdul-Latiff MAB, Md-Zain BM. Population census and age category character of Stump tailed macaque, *Macaca arctoides*, in Northern Peninsular Malaysia. Biodiversitas. 2019;**20**(9):2446-2452

[13] Jayaraj VK, Daud SHM, Azhar M-I, Sah SAM, Mokhtar SI, Abdullah MT. Diversity and conservation status of mammals in Wang Kelian State Park, Perlis. Malaysia. Check List. 2013;**9**(6): 1439-1448

[14] Malaivijitnond S, Hamada Y. A new record of stump-tailed macaques in Thailand and the sympatry with long-tailed macaques. The Natural

History Journal Chulalongkorn University. 2005;**5**(2):93-96

[15] Saaban S, Yazid AZ, Mustapa AR, Keliang C. (2016) In MPOC / SWD Human – Wildlife Conflict Workshop held on 22-23 November 2016 at FourPoints by Sheraton, Sandakan, Sabah. Available at http://mpoc.org.my/human-wildlife-conflict-in-peninsular-malaysia-current-status-and-overview-2/upload/Paper%202%20-%20PERHILITAN%20-%20Human-Wildlife%20Conflict%20in%20Peninsular%20Malaysia%20-%20Current%20Status%20and%20Overview.pdf

[16] Sabah Wildlife Department (2016) In MPOC / SWD Human – Wildlife Conflict Workshop held on 22-23 November 2016 at FourPoints by Sheraton, Sandakan, Sabah. Available at: http://www.mpoc.org.my/upload/Paper%201%20-%20Sabah%20Wildlife%20Department%20-%20Human-Wildlife%20Conflict%20in%20Sabah%20-%20Current%20Status%20&%20Overview.pdf

[17] Sadili A. Human-macaque conflict between tourists and long- tailed macaques in Kanching recreational forest [Masters thesis]. Rawang, Selangor, Malaysia: Universiti Putra Malaysia; 2016

[18] Hambali K, Ismail A, Zulkifli SZ, Md-Zain BM, Amir A. Human-macaque conflict and pest behaviors of long-tailed macaques (*Macacafascicularis*) in Kuala Selangor Nature Park. Tropical Natural History. 2012;**12**(2):189-205

[19] Osman NA, Abdul-Latiff MAB, Mohd-Ridwan AR, Yaakop S, Nor SM, Md-Zain BM. Diet composition of the wild stump-tailed macaque (Macaca arctoides) in Perlis State Park, Peninsular Malaysia, using a chloroplast tRNL DNA metabarcoding approach: A preliminary study. Animals (Basel). 2020;**10**(12):2215. DOI: 10.3390/ani10122215

[20] Morse SS, Mazet JA, Woolhouse M, Parrish CR, Carroll D, Karesh WB, et al. Prediction and prevention of the next pandemic zoonosis. Lancet. 2012; **380**(9857):1956-1965. DOI: 10.1016/S0140-6736(12)61684-5 PMID: 23200504; PMCID: PMC3712877

[21] Davies TJ, Pedersen AB. Phylogeny and geography predict pathogen community similarity in wild primates and humans. Proc Roy Soc B. 2008;**275**:1695-1701

[22] Anon. Bush meat widely sold in Sabah. In: The Star. Petaling Jaya, WP: Star Media Group Berhad; 2015 https://www.thestar.com.my/news/nation/2015/11/23/bush-meat-widely-sold-in-sabah/

[23] Lee TM, Sigouin A, Pinedo-Vasquez M, Nasi R. The harvest of wildlife for bushmeat and traditional medicine in East, South and Southeast Asia: Current knowledge base, challenges, opportunities and areas for future research. In: Occasional Paper 115. Bogor, Indonesia: CIFOR; 2014

[24] Then S. Trade in bushmeat and body parts depleting wildlife population in Sarawak. In: The Star. 2020 https://www.thestar.com.my/metro/metro-news/2020/12/01/trade-in-bushmeat-and-body-parts-depleting-wildlife-population-in-sarawak

[25] Yii MCK, Mohd-Azlan J. Wildlife Hunting and Utilization in Ulu Baleh, Sarawak, Malaysian Borneo. Ethnobiology Letters. 2020;**11**(1):76-84. DOI: 10.14237/ebl.11.1.2020.1647

[26] Shepherd C, Shepherd L. The poaching and trade of Malayan sun bears in Peninsular Malaysia. TRAFFIC Bulletin. 2010;**23**:49-52

[27] Ahmad T, Khan M, Mussa TH, Nasir S, Hui J, Bonilla-Aldana DK, et al. COVID-19: Zoonotic aspects. In: Travel Medicine and Infectious Disease. Advance online publication;

2020;**36**:101607. DOI: 10.1016/j.tmaid.2020.101607

[28] Corlett RT, Primack RB, Devictor V, Maas B, Goswami VR, Bates AE, et al. Impacts of the coronavirus pandemic on biodiversity conservation. Biological Conservation. 2020;**246**:108571

[29] Daly, N. (2020). Seven More Big Cats Test Positive for Coronavirus at Bronx Zoo. https://www.nationalgeographic.com/animals/2020/04/tiger-coronavirus-covid19-positive-testbronx-zoo/

[30] Kideghesho JR, Kimaro HS, Mayengo G, Kisingo AW. Can the Tanzania wildlife sector survive the COVID-19 pandemic? Tropical Conservation Science. 2021;**14**:1-18. DOI: 10.1177/19400829211012682

[31] Terraube J, Fernández-Llamazares Á. Strengthening protected areas to halt biodiversity loss and mitigate pandemic risks. Current Opinion in Environmental Sustainability. 2020;**46**:35-38. DOI: 10.1016/j.cosust.2020.08.014

[32] Thongyuan S, Sanyathitiseri P, Viriyarumpa S, Duangrasamee S, Boonkusol D, Tulayakul P. A study of bacterial contamination in feces of macaques in Lopburi Province, Thailand. International Journal of Infectious Diseases. 2016;**53**(Supplement):63. DOI: 10.1016/j.ijid.2016.11.159

[33] Mohamad MA. Identification of zoonotic pathogens in working pig-tailed macaques (*Macaca nemestrina*) In Kota Bharu &Bachok, Kelantan (Unpublished master's thesis). Malaysia: Universiti Malaysia Kelantan; 2021

[34] Burgos-Rodriguez AG. Zoonotic diseases of primates. Veterinary Clinics of Exotic Animals. 2011;**14**:557-575

[35] Ahmad T, Khan M, Haroon MTH, Nasir S, Hui J, Bonilla-Aldana DK, et al. COVID-19: Zoonotic aspects. Travel Med Infect Dis. 2020;**36**:101607. DOI: 10.1016/j.tmaid.2020.101607

[36] Sam IC, Chua CL, Rovie-Ryan JJ, Fu JY, Tong C, Sitam FT, et al. Chikungunya Virus in Macaques, Malaysia. Emerging Infectious Diseases. 2015;**21**(9):1683-1685. DOI: 10.3201/eid2109.150439

[37] Ain-Najwa MY, Yasmin AR, Arshad SS, Omar AR, Abu J, Kumar K, et al. Exposure to Zoonotic West Nile Virus in Long-Tailed Macaques and Bats in Peninsular Malaysia. Animals (Basel). 2020;**10**(12):2367. DOI: 10.3390/ani10122367 PMID: 33321964; PMCID: PMC7764493

[38] Woon Y, Lim M, Rashid TA, Thayan R, Chidambaram SK, Syed Abdul Rahin SS, et al. Zika virus infection in Malaysia: An epidemiological, clinical and virological analysis. BMC Infectious Diseases. 2019;**19**(152). DOI: 10.1186/s12879-019-3786-9

[39] Chua C, Chan Y, Andu E, Rovie-Ryan JJ, Sitam F, Verasahib K, et al. Little Evidence of Zika Virus Infection in Wild Long-tailed Macaques. Peninsular Malaysia. Emerging Infectious Diseases. 2019;**25**(2):374-376. DOI: 10.3201/eid2502.180258

[40] Lam SK, Chua KB, Hooi PS, Rahimah MA, Kumari S, Tharmaratnam M, et al. Chikungunya infection--an emerging disease in Malaysia. The Southeast Asian Journal of Tropical Medicine and Public Health. 2001;**32**(3):447-451 PMID: 11944696

[41] AbuBakar S, Sam IC, Wong PF, MatRahim N, Hooi PS, Roslan N. Reemergence of endemic Chikungunya, Malaysia. Emerging Infectious Diseases. 2007;**13**(1):147-149. DOI: 10.3201/eid1301.060617. PMID: 17370532; PMCID: PMC2725805

[42] Diallo M, Thonnon J, Traore-Lamizana M, Fontenille D. Vectors of chikungunya virus in Senegal: current data and transmission cycles.

The American Journal of Tropical Medicine and Hygiene. 1990;**60**:281-286

[43] Suhana O, Nazni WA, Apandi Y, Farah H, Lee HL, Sofian-Azirun M. Insight into the origin of chikungunya virus in Malaysian non-human primates via sequence analysis. Heliyon. 2019;**5**(12):e02682. DOI: 10.1016/j.heliyon.2019.e02682

[44] Cohen JI. Chapter 144 - Herpes B Virus. In: Bennett JE, Dolin R, Blaser MJ, editors. Mandell, Douglas, and Bennett's Principles and Practice of Infectious Diseases (Eighth Edition). New York: W.B. Saunders; 2015. pp. 1783-1786, ISBN 9781455748013. DOI: 10.1016/B978-1-4557-4801-3.00144-2

[45] Lee M, Rostal MK, Hughes T, Sitam F, Lee C, Japning J, et al. Macacine herpesvirus 1 in long-tailed Macaques, Malaysia, 2009-2011. Emerging Infectious Diseases. 2015;**21**(7):1107-1113. DOI: 10.3201/eid2107.140162

[46] Engel GA, Jones-Engel L, Schillaci MA, Suaryana KG, Putra A, Fuentes A, et al. Human exposure to herpesvirus B-seropositive macaques, Bali, Indonesia. Emerging Infectious Diseases. 2002 Aug;**8**(8):789-795. DOI: 10.3201/eid0808.010467 PMID: 12141963; PMCID: PMC3266706

[47] Elmore D, Eberle R. Monkey B virus (Cercopithecine herpesvirus 1). Comparative Medicine. 2008;**58**(1):11-21

[48] Bryan BL, Espana CD, Emmons RW, Vijayan N, Hoeprich PD. Recovery from encephalomyelitis caused by herpesvirus simiae: Report of a case. JAMA Internal Medicine. 1975;**135**(6):868-870

[49] Eberle R, Jones-Engel L. Questioning the extreme neurovirulence of monkey B virus (Macacinealphaherpesvirus 1). Advances in Virology. 2018;**2018**: 5248420, 17 pages. DOI: 10.1155/2018/5248420

[50] Jones-Engel L, May CC, Engel GA, Steinkraus KA, Schillaci MA, Fuentes, A....Linial, M. L. Diverse contexts of zoonotic transmission of simian foamy viruses in Asia. Emerging Infectious Diseases. 2008;**14**(8):1200-1208. DOI: 10.3201/eid1408.071430

[51] Jones-Engel L, Steinkraus KA, Murray SM, Engel GA, Grant R, Aggimarangsee N, et al. Sensitive assays for simian foamy viruses reveal a high prevalence of infection in commensal, free-ranging Asian monkeys. Journal of Virology. 2007;**81**(14):7330-7337. DOI: 10.1128/JVI.00343-07

[52] Hussin N, Lim YAL, Goh PP, et al. Updates on malaria incidence and profile in Malaysia from 2013 to 2017. Malaria Journal. 2020;**19**:55. DOI: 10.1186/s12936-020-3135-x

[53] Yusof R, Lau YL, Mahmud R, Fong MY, Jelip J, Ngian HU, et al. High proportion of knowlesi malaria in recent malaria cases in Malaysia. Malaria Journal. 2014 May;**3**(13):168. DOI: 10.1186/1475-2875-13-168 PMID: 24886266; PMCID: PMC4016780

[54] Singh B, Daneshvar C. Plasmodium knowlesi malaria in Malaysia. The Medical Journal of Malaysia. 2010;**65**(3):166-172 PMID: 21939162

[55] Rahim MAFA, Munajat MB, Idris ZM. Malaria distribution and performance of malaria diagnostic methods in Malaysia (1980-2019): A systematic review. Malaria Journal. 2020;**19**:395. DOI: 10.1186/s12936-020-03470-8

[56] Kurniawati DA, Suwanti LT, Lastuti NDR, Kusdarto S, Suprihati E, Mufasirin M, et al. Zoonotic potential of gastrointestinal parasite in long-tailed Macaque Macaca fascicularis at Baluran National Park, Situbondo, East Java. Indonesia. Aceh Journal of Animal Science. 2020;**5**(1):47-56. DOI: 10.13170/ajas.5.1.15397

[57] Damrongsukij P, Doemlim P, Kusolsongkhrokul R, Tanee T, Petcharat P, Siriporn B, et al. One health approach of melioidosis and gastrointestinal parasitic infections from *Macaca fascicularis* to human at Kosumpee Forest Park, MahaSarakham, Thailand. Infect Drug Resist. 2021; **15**(14):2213-2223. DOI: 10.2147/IDR. S299797 PMID: 34163186; PMCID: PMC8214530

[58] Adrus M, Zainudin R, Ahamad M, Jayasilan MA, Abdullah MT. Gastrointestinal parasites of zoonotic importance observed in the wild, urban, and captive populations of non-human primates in Malaysia. Journal of Medical Primatology. 2019;**48**(1):22-31. DOI: 10.1111/jmp.12389 Epub 2018 Oct 29. PMID: 30370934

[59] Choong SS, Mimi Armiladiana M, Ruhil HH, Peng TL. Prevalence of parasites in working pig-tailed Macaques (*Macaca nemestrina*) in Kelantan, Malaysia. J Med Primatol. 2019;**48**(4):207-210. DOI: 10.1111/jmp.12416

[60] Karuppannan KV, Saaban S, Firdaus Ariff AR, Mustapa AR. Non-surgical castration in controling long tailed macaque (Macaca fascicularis) population by Deparatment of Wildlife and National Parks (DWNP) Peninsular Malaysia. Malaysian Journal of Veterinary Research. 2013;**4**(1):33-36

[61] Takeshita RSC, Huffman MA, Kinoshita K, Bercovitch FB. Effect of castration on social behavior and hormones in male Japanese macaques (*Macaca fuscata*). Physiology & Behavior. 2017;**181**:43-50

[62] Wang Q, Kessler MJ, Kensler TB, Dechow PC. The mandibles of castrated male rhesus macaques (*Macaca mulatta*): The effects of orchidectomy on bone and teeth. American Journal of Physical Anthropology. 2016;**159**(1):31-51. DOI: 10.1002/ajpa.22833

[63] Kumar V, Kumar V. Clinical evaluation of laparoscopic sterilization techniques in female Rhesus Macaques (*Macaca mulatta*). Archives of Veterinary Science. 2012;**17**(3):20-26

[64] Yu PH, Weng CC, Kuo HC, Chi CH. Evaluation of endoscopic salpingectomy for sterilization of female Formosan macaques (Macaca cyclopis). American Journal of Primatology. 2014;**77**(4):359-367. DOI: 10.1002/ajp.22354

[65] Lappan S, Ruppert N. Primate research and conservation in Malaysia. CAB Reviews. 2018;**14**(4):1-10

[66] Holzner A, Balasubramaniam KN, Weiß BM, et al. Oil palm cultivation critically affects sociality in a threatened Malaysian primate. Scientific Reports. 2021;**11**:10353. DOI: 10.1038/s41598-021-89783-3

[67] Dzulhelmi MN, Suriyanti S, Manickam S. Population, behaviour and conservation status of long-tailed macaque, *Macaca fascicularis* and southern pig-tailed macaque, *Macaca nemestrina* in Paya Bakau Park, Perak, Malaysia. The Journal of Animal & Plant Sciences. 2019;**29**(2):611-618

Chapter 4

Diseases as Impediments to Livestock Production and Wildlife Conservation Goals

Y.J. Atuman, C.A. Kudi, P.A. Abdu, O.O. Okubanjo and A. Abubakar

Abstract

Disease outbreaks, epidemics or pandemics have been of importance for human and animal health worldwide and sparked enormous public interest. These outbreaks might be caused by known endemic pathogens or by emerging or re-emerging pathogens. Wildlife are the major reservoirs and responsible for most of these outbreaks. They play significant role in the transmission of several livestock diseases and pathogen spill-over may occur in complex socio-ecological systems at the wildlife-domestic animal interface which have been seldom studied. Interspecific pathogen spill-over at the wildlife-livestock interface have been of growing concern in the scientific community over the past years due to their impact on wildlife, livestock and human health. In this section the epidemiology of some viral infections (Foot and Mouth Disease and rabies), bacterial infections (Tuberculosis and brucellosis) and parasites (haemo and endo-parasites) at the wildlife-livestock interface and potential impacts to livestock production and conservation goal is described.

Keywords: wildlife, livestock, diseases, conservation goals, public health

1. Introduction

Disease transmission is one of the major obstacles for the survival of wildlife and livestock in sub-Saharan Africa [1–3]. Movements of both human and animals occur across boundaries that separate farming and wildlife activities, and this results into direct and/or indirect contact between wild and domestic animals, potentially leading to diseases transmission. This is considered the major challenge of people living in areas where wildlife and domestic animals frequently come into contact. Movement of livestock and wildlife across the boundaries of protected areas provides ideal condition for the transmission of pathogens in either direction from wildlife to livestock and vice-versa [2]. At the edges of protected areas, the occurrence of diseases in wildlife and livestock can be categorized into wildlife-maintained diseases, in which wildlife are resistant or silent carriers of the infection and the second is multi-species diseases, which have serious outcomes for both wildlife and livestock [3]. The role of wildlife as disease reservoir is well established, their involvement in the occurrence of vector-borne infections in domestic animals has gained renewed interest as emerging and re-emerging infections occurring worldwide at an increasing rate [4–6]. Wildlife may also act as introductory or transporting hosts as a result of their movement to new regions, e.g. Rift Valley Fever virus and African horse sickness virus. They may also act as amplifying hosts with spill-over to domestic animals, e.g. African horse sickness [7]. Other important wildlife- maintained diseases include Foot and Mouth Disease (FMD), Malignant Catarrhal Fever, trypanosomiasis, theileriosis, ehrlichiosis, lumpy skin disease [2]. Multi-species diseases mostly have fatal outcome in both wildlife and domestic animals and are frequently zoonotic, i.e. transmitted among wildlife, livestock and humans. They include anthrax, rabies, brucellosis and bovine tuberculosis [2, 3]. Pathogens from wildlife can be modified to cause severe outbreaks in humans and animals due to gradual change of many related factors like environment and ecological factors [8, 9]. These factors include heavy rainfall, change in the onset of the rainy season (and thus an alteration of the interplay between rainfalls and reproduction of vectors and reservoir hosts) and changes in wind patterns [9, 10]. Changes in biodiversity, land use and climate in particular have been identified as the major drivers of disease emergence [4, 6, 11]. Loss of biodiversity, alterations of natural habitats as well as land use and cover changes are linked to emerging of 50% of zoonoses [4]. Elevation of ambient temperature might facilitate the proliferation of insects like mosquitoes [9] which may lead to increase in prevalence of some insect-borne livestock diseases such as Rift Valley Fever (RVF) and African Horse Sickness (AHS) [9, 12]. Most domesticated and wild animals perform optimally at temperatures between 10 and 30°C [13]. At temperatures above 30°C, most domestic and wild animals would reduce their feed intake by 3–5% for each unit temperature increase [14]. Changes in temperature also trigger the secretion of stress hormones which suppress immunological responses including the function of the white blood cells, tipping the host-pathogen interactions in favor of the pathogen [14].

2. Role of livestock in disease outbreaks at the edges of protected areas

Land use, human and animal movements and grazing by livestock have been considered as major factors responsible for the spread of diseases from domestic animals to wild populations and vice versa [15]. The growth in human population has resulted in greater use of land thereby establishing human settlements around the protected areas and bringing livestock closer to wild populations [11, 15]. Also, increase in the demand for protein has led to expansion of livestock farms and ranches thereby increasing wildlife-livestock interface [15]. Other risk factors for pathogen spread at the wild-domestic animal interface include long distance live animal transportation, live animal markets, and building of dams, bush meat consumption and habitat destruction [16]. These have resulted in the emergence and spread of pathogens that are of economic and public health concern [11, 17–19, 34]. Poor rural households in developing countries often survive on a combination of on-farm and off-farm activities [3]. Use of natural resources from the game parks acts as coping strategy during periods of socio-economic instability or environmental shocks [20]. Wild food products play a major role in societies that experience seasonal food shortages [3], while access to water and grazing resources during the droughts is a matter of survival for livestock in the semi-arid ecosystem of sub-Saharan Africa [21–23]. Competition with wildlife for these scarce resources exacerbates such scarcities for people living at the fringes of protected areas [21, 22, 24]. At the same time, protected areas provide an alternative source of grazing, and thus a 'buffer' during these difficult periods thereby providing opportunities for disease spread from livestock to wildlife through direct and/or indirect contact [3]. Most protected areas bordering human settlements and farmlands in Africa are subjected to growing human-wildlife conflicts and increasing incidences of pathogens transmitted between domestic and wild animals [25, 26]. The cases in point are Yankari Game Reserve (YGR) in Nigeria [25] and Ngorongoro Conservation Area (NCA) in Tanzania [27]. Diseases can be easily transmitted from domestic livestock populations into wild animal populations at the edge of protected areas and whenever a disease condition is established in a wild population, control measures in both the wildlife and livestock populations become much problematic. It has been shown that an infected wildlife reservoir that interacts with livestock causes frequent herd breakdown and substantial economic losses to agricultural sector [28].

3. Role of wildlife in disease outbreaks at the edges of protected areas

Throughout history, wildlife has been an important source of infectious diseases transmissible to humans and domestic animals in virtually all continents [29, 30]. Disease outbreaks might be caused by known endemic pathogens or by emerging or re-emerging pathogens [31]. Wildlife are the major reservoirs and responsible for 70% of these outbreaks [30, 32, 33]. Three-quarters of all emerging infectious diseases of humans are zoonotic, most of which are of wildlife origin, with an increasing incidence since the 1940s [11, 34]. Wildlife play significant role in transmission of several livestock diseases and interspecific viral pathogen spill-over at the wildlife-livestock interface and have been of growing concern in the scientific community over the past years due to their impact on wildlife, livestock and human health [35, 36]. Most infectious diseases associated with wildlife reservoir are typically caused by various bacteria, viruses, and parasites, whereas fungi are of negligible importance [37]. The current rapid ecological changes in the world have negative impact on pathogenic organisms, their vectors and hosts which are equally capable of rapid change [38]. Some of these pathogens may cause significant disease in wild species, but in other cases the wild animals may serve as reservoirs for pathogens which do not induce overt illness in their wild hosts [38]. Wildlife have been recognized as reservoirs of infectious diseases; it is advocated that interdisciplinary wildlife disease surveillance system using modern laboratory techniques be utilized for effective control of disease spread at the edge of protected areas.

4. Epidemiology of some wildlife-related viral infections at the edges of protected areas

The role of African wildlife in the occurrence of infectious diseases in domestic animals has gained renewed interest as emerging and re-emerging infections increase worldwide [7, 27, 39]. Viral diseases originating from wild animals are widely considered the major threats to public health and transmission of such viral pathogens from wildlife to other domestic animals and humans remains an important scientific challenge hampered by pathogen detection limitations in wild species [40]. Viral pathogens spill over at the edge of protected areas particularly in remote communities/areas is mostly under-reported and the disease events occur undetected [41]. Characterizing epidemiologic features of viral transmission at the wildlife-livestock interface has also revealed a number of high-risk human activities that have enabled virus spill-over in situations that facilitate close contact among wild and domestic animals and people. Domestication of animals, human encroachment into wildlife habitats and hunting of wild animals are key anthropogenic activities driving viral disease emergence at the global scale and in most instances these activities have contributed to wildlife population declines and extinction [42]. It is recommended that in-depth investigation of the epidemiology of viral infections at the edges of protected areas should be intensified to assess the risks of viral disease emergence for effective disease prevention and control measures. Foot and Mouth Disease, Rift Valley Fever and Rabies, discussed hereunder, are three of the viral wildlife-related viral infections which are common at the edges of wildlife protected areas in Sub-Saharan Africa.

4.1 Foot and Mouth Disease (FMD)

Foot and Mouth Disease is one of the most economically important transboundary diseases of animals in the world and it is extremely difficult to control due to involvement of wildlife as reservoir host [43, 44]. In wildlife, pathogenesis of FMD varies from a completely inapparent to a rare acutely lethal infection and this makes diagnosis difficult [36, 44, 45]. The transmission of FMDV in sub-Saharan Africa is mainly driven by two epidemiological cycles: one in which wildlife plays a significant role in maintaining and spreading the disease to other susceptible wild and/or domestic ruminants [46]. Whilst with the second cycle, the virus is solely transmitted within domestic populations and hence is independent of wildlife [36].

FMD is currently found in limited areas (small pockets/regions) of Europe and also in Africa, Middle East, and Asia and has contributed significantly to decline of wildlife and livestock populations in those regions [2, 26]. Presence of antibodies against the FMDV in several wildlife species has been documented in studies conducted in different African countries including South Africa, Nigeria and Tanzania [28, 47–49]. Fifty-one FMD outbreaks occurred involving buffalo (*Syncerus caffer*), impala (*Aepyceros melampus*) and elephant (*Loxodonta africana*) in the Kruger National Park (KNP) and adjacent areas of South Africa between 1970s and 2000s, of which 16 were SAT1, 31 were SAT2, 4 were SAT3 and 3 were not serotyped and the outbreaks spilled over to other species of wild animals and livestock [50]. Previous findings also established that monitoring of FMD in wild animals like impala and buffalo is important because they can serve as a source of infection for livestock [48, 51, 52]. Recent studies in Nigeria showed that, presence of wildlife population along the protected areas where cloven hoofed species come into contact with livestock contributed to high FMDV seropositivity in livestock due to spill-over of FMDV from wildlife to livestock [49, 53]. Similar studies in Tanzania revealed a high exposure patterns in buffalo populations in Ngorongoro Conservation Area, Serengeti, Katavi and Tarangire National Parks [54]. The results suggest that wildlife could play an important role in the epidemiology of FMD.

4.2 Rift Valley fever virus (RVF)

This is an emerging zoonotic disease of public and animal health concern [55–59]. It is endemic in many countries in Sub-Saharan Africa and is responsible for severe outbreaks in livestock characterized by a sudden onset of abortions and high neonatal mortality which results in significant economic losses [55, 60]. The virus was first identified in 1931 during an investigation into an epidemic among sheep on a farm in the Rift Valley of Kenya [61]. Many African wildlife species have tested positive for antibody against RVFV, including: topi (*Damaliscus korrigum*); red-fronted gazelle (*Eudorcas rufifrons*); dama gazelle (*Nanger dama*); scimitar-horned oryx (*Oryx dammah*); common reedbuck (*Redunca redunca*); African buffalo (*Synceruscaffer*); Dorcas gazelle (*Gazella dorcas*); Thomson's gazelle (*Gazella thomsonii*); gerenuk (*Litocraniuswalleri*); lesser kudu (*Tragelaphus strepsiceros*); impala (*Aepyceros melampus*); sable antelope (*Hippotragus niger*); waterbuck (*Kobus ellipsiprymnus*); warthog (*Phacochoerus aethiopicus*); African bush elephant (*Loxodonta africana*); giraffe (*Giraffa camelopardalis*); Burchell's zebra (*Equus burchellii*); and black rhinoceros (*Diceros bicornis*) [62, 63]. Although serological evidence suggests that a large number of African wildlife species might play a role in the epidemiology of RVF, their possible role in the cryptic maintenance of the virus is poorly understood [62, 64].

4.3 Rabies

Rabies is endemic in several African countries. It is a vaccine-preventable fatal viral disease of human and mammals [65] and is responsible for numerous human deaths. Dogs are the major reservoirs and primary source of rabies virus transmission to both humans and other animals in most developing countries [37]. Other domestic and wild animals are typically infected through secondary transmission of rabies virus variants that are maintained by dogs or in some cases variants of rabies virus maintained by wild carnivore hosts [66]. Rabies virus is widely distributed and affects various animals [67, 68]. In North America and Europe, wildlife species have replaced dogs as the most important reservoirs of rabies and new viral etiologic agents continue to emerge [68–70]. However, in Nigeria and other African countries, there has been a low but consistent number of positive cases from wildlife species which were reported over the years [26, 71–78]. Human activities at the fringes of game reserves in most parts of Africa where domestic dogs are used for hunting purposes and provision of security to livestock and farm crops against problem wildlife species [25, 79]. Majority of the dogs found at the fringes of game reserves are not adequately taken care of in terms of vaccination, feeding and provision of shelter and mostly roam about to scavenge for food creating opportunities for contact with wildlife [25]. Rabies antigens detected in jackals and mongooses in Bauchi State, Nigeria were associated with spill-over from dogs at the fringes of Yankari Game Reserve (YGR) and Sumu Wildlife Park (SWP) [25]. It is also important to note that while dogs' impacts to wildlife is likely to occur at individual level the results may still have important implications for wildlife populations. RABV has affected mammalian species at an elevated extinction risk from multiple taxonomic orders (Carnivora, Chiroptera, Primates and Proboscidea). For certain species, such as the African wild dog (*Lycaon pictus*) and Ethiopian wolf (*Canis simensis*), RABV outbreaks have led directly to severe decreases in population size and local endangerment or extinction [80]. For other species of conservationconcern, occasional rabies cases may contribute to overall population declines in conjunction with other pressures, such as habitat fragmentation, decreased food availability and illegal killing. Both endemic strains of domestic dog RABV and sylvatic RABV strains have been implicated in the infection of many of these taxa, which is important information in terms of conservation and control [77]. Once stable control of RABV is achieved in domestic dogs, remaining rabies threats to wildlife conservation can be addressed more effectively.

5. Epidemiology of some wildlife-related bacterial infections at the fringes of game reserves

Transmission of bacterial disease that can be spread between wild and livestock animals at the edge of protected areas occurs directly via physical contact and or indirectly via environmental exposure, with devastating consequences for human and animal health, as well as pastoral economies [81]. The wildlife–livestock interface, is recognized overtime as the driver for inter-species bacterial pathogens transmission between animals and subsequent potential spill-over to humans, whereas habitat loss which is associated with altering the abundance, richness of the wildlife ecosystem and movement patterns of the wildlife species can directly or indirectly influence wildlife-livestock interfaces [81]. Bovine tuberculosis and Brucellosis are representative of the bacterial diseases transmitted between wildlife and domestic species.

5.1 Bovine tuberculosis (bTB)

Bovine tuberculosis (bTB) is caused by *Mycobacterium bovis* (*M. bovis*). It is a chronic, infectious and contagious disease of livestock, wildlife and humans [82, 83]. The disease is an important public health concern worldwide, especially in developing countries, due to deficiency in preventive and/or control measures [83, 84]. Members of the closely related phylogenic grouping of *Mycobacterium* known collectively as the *Mycobacterium tuberculosis* complex may cause tuberculosis in a range of species including human. Some members of this group are predominately found in human (*M. tuberculosis*, *M. africanum*, *M. canetti*) or rodent pathogens (*M. microti*), whereas, others have wide host spectrum (*M. bovis*, *M. caprae*) [85]. The respiratory route is accepted as the primary method of infection spread in all species. However, it is clear that there are other less common methods of spread such as oral [85]. African Wildlife were said to have been infected with Bovine tuberculosis from infected imported cattle and the disease is now endemic in wildlife [86, 87]. In Ireland and Great Britain, badgers (*Taxidea taxus*) maintain the infection, whereas the brushtail possum (*Trichosurus vulpecula*) constitutes a main wildlife reservoir in New Zealand. In parts of Michigan, bovine tuberculosis is endemic among white-tailed deer (*Odocoileus virginianus*), whereas in Europe, both wild boars and various deer species can be a reservoir of the pathogen [88]. The natural movement of these reservoir animals increases the spread of the disease to domestic animals and thereby posing a major public health impact [88].

Bovine tuberculosis in Antelopes has been reported in Nigeria [89]. This is not surprising, as there had been evidence of exposure in similar antelope species in other countries [89, 90]. In Africa, both wildlife and livestock share the rich pasture resources available at the edges of protected areas, thus creating an ideal condition for transmission of Tuberculosis through contact with infected animals. Also, poaching and slaughter of cattle for meat could contribute in the spread of infection of these disease among animals and humans far away from the game reserve from infected tissues and contamination of the environment. Pastoralists and agro-pastoralists are considered high risk groups for contracting bTB and brucellosis due to their close association with livestock and diets rich in animal products [91]. Both diseases have been reported in many wildlife, livestock and human interfaces in Africa [92–94]. Bovine tuberculosis's main route of transmission is through aerosol, hence, people engaging in tourism and ecotourism to the game reserves are also at risk. This is more so because of the human population growth and associated changes, as well as competition for grazing lands, have made wildlife-livestock disease transmission more likely by reducing the spatial separation between livestock operations and wildlife habitat [95].

5.2 Brucellosis

The disease primarily affects domestic animals including cattle, pigs, sheep, goats and occasionally horses. In wildlife, the prevalence could be low but there is always a clear epidemiological link between wildlife and domestic animals [96]. *Brucella* organism was first described as far back 1887 as *Micrococcus melitensis* [96]. The causative organism was later renamed *Brucella melitensis* and has been rated by WHO as one of the most important zoonoses, as it is very pathogenic to humans, causing the disease known as Malta fever (also known as Mediterranean or undulating fever) [96]. The ability of Brucella organisms to be transmitted, rapidly and efficiently, over long distances and the socio-economic impact of the disease in both human and livestock necessitated the increased awareness of the existence of the disease worldwide [97]. A lot of wild animals are also affected with Brucellosis [98, 99] and the disease is increasingly important in wildlife conservation, particularly when endangered species are involved [99].

6. Epidemiology of wildlife related parasitic infections at the fringes of game reserves

Parasites play an important role in the dynamics of wildlife populations [100]. They can cause substantial losses in production or even acute clinical signs and death [101]. There is abundant evidence of parasitic infections in wildlife worldwide and studies have demonstrated that they may be carriers of gastrointestinal parasites [102–106], ectoparasites [107, 108] and haemoparasites [109–111]. Many wildlife species are capable of living with high parasite loads without any apparent ill-effect on their health [112]. The impact of spill-over of human and livestock parasites to naïve species of wildlife and spill-back from wildlife is another emerging threat of potential public health and economic significance to humans, wildlife and livestock [113]. Ticks suck blood of their hosts resulting into severe anemia, loss of production, weakness and immunosuppression [114, 115] as well as damages to hides and skin leading to significant financial losses to livestock farmers [116]. Production losses due to ticks and tick-borne diseases around the globe were put at US$ 13.9 to US$ 18.7 billion annually [117]. With the establishment of zoos and conservation areas [118, 119], wildlife-livestock interface is found in proximity to many protected areas in Africa [26].

Parasitic infections transmitted between livestock and wildlife pose a significant risk to wildlife conservation efforts and constrain livestock productivity in tropical regions of the world [120]. The emergence of infectious diseases with zoonotic potential has dominated investigations and commentary on wildlife pathogens [26, 121]. In terms of conservation, it is unfortunate that by doing so, not only have studies on the biodiversity and ecology of wildlife parasites been neglected, but control efforts have also been hampered [122]. In Africa, biodiversity conservation and the expansion of livestock production have increased the risk of transmitting vector-borne infections between wildlife and livestock [7]. In addition to the physical injury caused by parasites, some serve as hosts of many viral, rickettsial, bacterial and protozoan diseases [123–125]. In Nigeria, wildlife conservation areas such as Yankari Game Reserve (YGR) and Sumu Wildlife Park (SWP) are natural heritage and means of generating revenue [126]; and parasitic infections may constrain the health of the variety of wildlife species in these conservation areas. There is abundant evidence of haemoparsitic infections in wildlife worldwide, in some circumstances displaying high prevalence with some of them serving as reservoirs for the haemoparasites [125]. There are wide ranges of potential vectors that may allow these parasites to maintain endemic sylvatic life-cycles in their geographical distribution area [127, 128]. This could potentially lead to the transmission of infection to domestic species, especially in peri-urban and urban environments [125]. Wildlife species are reservoirs of parasitic infections and have the ability to expand their geographical ranges, thus increasing intra- and interspecies contact risk with domestic animals and spread of infective parasites [125, 129]. However, a high prevalence of infection alone does not demonstrate that the species in question acts as a reservoir, some wildlife are not abundant, and probably unable to maintain a pathogen in the absence of domestic species reservoirs [125]. The epidemiology of multilocular echinococcosis, caused by the small tapeworm *Echinococcus multilocularis*, has also been influenced by the translocation of animals where the main hosts especially foxes, the intermediate hosts small rodents and human accidental hosts were found to be positive with *E. multilocularis* [129–131]. Wildlife species were found to be infested with various ticks as *Amblyoma, Rhipicephalus, Hyaloma* and *Boophilus* genus during study in Nigeria and other African countries and such tick infestation has been suggested as the cause of mortality in several ungulate species [132–134].

Wildlife management systems in most game reserves and game parks which subjects them to continuous challenge of vectors, scarcities of feeds and stress from environmental and climate variations coupled with illegal livestock and human activities are compounding factors to efforts at controlling parasitic infections in those areas.

7. Conclusion

Disease outbreaks have affected human and animal health throughout times, and wildlife has always played a role. The ecological changes influencing the epidemiology of wildlife-related viral, bacterial and parasitic infections can be of natural or anthropogenic origin. These include, but are not limited to, human population expansion and encroachment, reforestation and other habitat changes, pollution, and environmental and climatic changes. The movement of pathogens, vectors, and domestic animals including humans is another factor influencing the epidemiology of wildlife-related disease (viral, bacteria and parasitic) outbreaks. Such movements are commonly encountered at the edges of protected areas due to availability of rich resources and bring about interactions at the wild/domestic animals/humans' interfaces with conflicts and potential for pathogen spread at the interface. These are emerging threats to wildlife conservation goals and livestock and human health. It is suggested that preventive measures should be geared towards improved disease surveillance among domestic and wild animals at the edges of protected areas using improved diagnostic techniques, vector control and implementation of restrictions on anthropogenic animal movement, concomitant with public enlightenment campaign and behavioral change. More so, collaborative, multisectoral, and transdisciplinary approach to surveillance and control of emerging and re-emerging diseases at the edge of protected areas at local, regional, national, and global levels should be intensified.

Author details

Y.J. Atuman[1]*, C.A. Kudi[2], P.A. Abdu[2], O.O. Okubanjo[3] and A. Abubakar[4]

1 National Veterinary Research Institute Vom Outstation Laboratory, Bauchi, Bauchi State, Nigeria

2 Department of Veterinary Medicine, Faculty of Veterinary Medicine, Ahmadu Bello University, Zaria, Nigeria

3 Department of Veterinary Parasitology and Entomology, Ahmadu Bello University, Zaria, Nigeria

4 Force Animal Branch Department, Nigeria Police Force Headquarters, Abuja, Nigeria

*Address all correspondence to: atumanyj@yahoo.com

References

[1] Bourn D, Blench R. Can Livestock and Wildlife Co-exist? An Interdisciplinary Approach. London: ODI; 1999. pp. 83-121

[2] Bengis R, Kock R, Fischer J. Infectious animal diseases: The wildlife-livestock interface. Revue Scientifique et Technique-Office International des Epizooties. 2002;**21**(1):53-66

[3] de GarineWichatitsky M., Frits, H Chaminuka, P., Caron, E., Guerbois, C., Pfukenyi, D.M., matema C., Jori, F and Murwira, A (2013). Consequences of animals crossing the edges of transfrontier parks. In Transfrontier Conservation Areas by Anderson, J.A., de Garine-Wichatitsky M., Cumming, D.H.M., Dzingirai, V. and Giller, K.E. Pp138-162

[4] Loh EH, Zambrana-Torrelio C, Olival KJ, Bogich TL, Johnson CK, Mazet JAK, et al. Targeting transmission pathways for emerging zoonotic disease surveillance and control. Vector Borne and Zoonotic Diseases. 2015;**15**(7):432-437

[5] Armstrong, F., Capon, A. and McFarlane, R. (2020). Coronavirus is a wake-up call: Our war with the environment is leading to pandemics. The Conversation, Available at https://theconversation.com/coronavirus-is-a-wake-up-call-our-war-with-the-environment-is-leading-to-pandemics-135023

[6] WWF (2020). The Loss of Nature and Rise of Pandemics: Protecting Human and Planetary Health. WWF International, Gland, Switzerland. 19pp

[7] Van Vuuren, M. Penzhorn, B. L. (2015). Geographic range of vector-borne infections and their vectors: the role of African wildlife. Pp 34-76.

[8] Martin C, Pastorate PP, Brucher B, Humblet M, Saegerman C. A survey of the transmission of infectious diseases/infections between wild and domestic ungulates in Europe. Veterinary Research. 2011;**42**:70

[9] Oluwayelu D. Potential impact of climate and environmental changes on Occurrenceand transmission of Arboviral diseases of livestock in Nigeria. Nigerian Veterinary Journal. 2014;**35**(3):1038-1052

[10] Sellers RF, Maarouf AR. Possible introduction of epizootic hemorrhagic disease of deer virus (serotype 2) and bluetongue virus (serotype 11) into British Columbia in 1987 and 1988 by infected Culicoides carried on the wind. Canadian Journal of Veterinary Research. 1991;**55**(4):367

[11] Jones KE, Patel NG, Levy MA, Storeygard A, Balk D, Gittleman JL, et al. Global trends in emerging infectious diseases. Nature. 2008;**451**(7181):990-993

[12] Gray JS. *Ixodes ricinus* seasonal activity: Implications of global warming indicated by revisiting tick and weather data. International Journal of Medical Microbiology. 2008;**298**:19-24

[13] Das R, Sailo L, Verma N, Bharti P, andSaikia J. Impact of heat stress on health and performance of dairy animals: A review. Veterinary World. 2016;**9**(3):260

[14] Dittmar J, Janssen H, Kuske A, Kurtz J, andScharsack JP. Heat and immunity: An experimental heat wave alters immune functions in three-spined sticklebacks (Gasterosteus aculeatus). Journal of Animal Ecology. 2014;**83**(4):744-757

[15] Kock R, Kebkiba B, Heinonen R, andBedane B. Wildlife and pastoral

society—Shifting paradigms in disease control. Annals of the New York Academy of Sciences. 2002;**969**(1):24-33

[16] MacLachlan N, and Dubovi E. Poxviridae. Fenner's Veterinary Virology. 4th ed. London: Academic; 2011. pp. 151-165

[17] Calisher CH, Childs JE, Field HE, Holmes KV, andSchountz T. Bats: Important reservoir hosts of emerging viruses. Clinical Microbiology Reviews. 2006;**19**(3):531-545

[18] Kaur T, Singh J, Tong S, Humphrey C, Clevenger D, Tan W, et al. Descriptive epidemiology of fatal respiratory outbreaks and detection of a human-related metapneumovirus in wild chimpanzees (Pan troglodytes) at Mahale Mountains National Park, Western Tanzania. American Journal of Primatology. 2008;**70**(8):755-765

[19] Smith DR, Holbrook MR, andGowen BB. Animal models of viral hemorrhagic fever. Antiviral Research. 2014;**112**:59-79

[20] Paumgarten F. The role of non-timber forest products as safety-nets: A review of evidence with focus on South Africa. GeoJournal. 2006;**64**(3):189-197

[21] de Garine-Wichatitsky, Michel., Y, SOUBEYRAN., Maillard, Daniel. and Duncan, Patrick. (2005). The diets of introduced rusa deer (*Cervus timorensis* russa) in a native sclerophyll forest and a native rainforest of New Caledonia. New Zealand Journal of Zoology. 32. 117-126. 10.1080/03014223.2005.9518403.

[22] Blench R. Drought Preparedness in Zimbabwe. Consultant Report to DFID Project Lessons for Drought Contingency Planning for Pastoral and Agro-Pastoral Areas. Chatham,UK: *Natural Resources Institute*; 1999. pp. 1-13

[23] Turner MD, Schlecht E. Livestock mobility in sub-Saharan Africa: Acritical review pastoralism: *Research*. Policy and Practice. 2019;**9**(13)

[24] Tchamba M. Conflict between people and protected areas within the Bénoué wildlife conservation area, North Cameroon. Oryx. 2003;**37**:72-79. DOI: 10.1017/S0030605303000140

[25] Atuman YJ, Adawa DAY, Audu S, Paul MP, andBankole OA. Potential Risks for Rabies Spill-Over from Apparently Healthy Dogs to Wildlife in Bauchi State, Nigeria. Journal of Veterinary Advances. 2014;**4**(4):493-498

[26] Daszak P, Cunningham AA, andHyatt AD. Emerging infectious diseases of wildlife--threats to biodiversity and human health. Science. 2000;**287**(5452):443-451

[27] Rwambo P, Grootenhuis JG, Demartini J, Mkumbo S. Animal Disease Risk in the Wildlife/Livestock Interface in the Ngorongoro Conservation Area of Tanzania, Report to Global Livestock Collaborative Research Support Program. Davis, California: University of California; 1999

[28] McDermott J, Arimi SM. Brucellosis in sub-Saharan Africa: Epidemiology, control and impact. Vet Microbiol. 2002;**90**(1-4):111-134. DOI: 10.1016/s0378-1135(02)00249-3

[29] Blancou J. History of the Surveillance and Control of Transmissible Animal Diseases. Paris: Office International des Epizooties; 2003. pp. 234-242

[30] Kruse H. Wildlife as source of zoonotic infections. Emerging Infectious Disease Journal. 2004;**10**(12):102-112

[31] Diederich S, Gedvilaite A, Zvirbliene A, Kazaks A, Sasnauskas K, Johnson N, et al. Virus-like particles: A versatile tool for basic and applied research on emerging and reemerging

viruses. In: Viral Nanotechnology. CRC Press; 2015. pp. 137-160

[32] Weiss, S., Nowak, K., Fahr, J., Wibbelt, G., Mombouli, J.V., Parra, H.J. and Leendertz, F. (2012). Henipavirus-related sequences in fruit bat bushmeat, Republic of Congo. pp 73-98

[33] Pernet O, Schneider BS, Beaty SM, LeBreton M, Yun TE, Park A, et al. Evidence for henipavirus spillover into human populations in Africa. Nature Communications. 2014;**5**

[34] Taylor LH, Latham SM, andMark E. Risk factors for human disease emergence. Philosophical Transactions of the Royal Society of London B: Biological Sciences. 2001;**356**(1411):983-989

[35] Karesh WB, Cook RA, Bennett EL, Newcomb J. Wildlife trade and global disease emergence. Emerging Infectious Diseases. 2005;**11**(7):1000

[36] Miguel E, Grosbois V, Caron A, Boulinier T, Fritz H, Cornélis D, et al. Contacts and foot and mouth disease transmission from wild to domestic bovines in Africa. Ecosphere. 2013;**4**(4):1-32

[37] World Health Organization. Consultation on public health and animal Transmissable spongiform Encephalopathies. Epidemiology, risk and research requirements. WHO/CDS/CSR/APH/2000.2. In: Geneva. 1999

[38] Williams E, Yuill T, Artois M, Fischer J, andHaigh S. Emerging infectious diseases in wildlife. Revue Scientifique et Technique-Office International des Epizooties. 2002;**21**(1):139-158

[39] Wiethoelter AK, Beltrán-Alcrudo D, Kock R, Mor SM. Global trends in infectious diseases at the wildlife–livestock interface. PNAS. 2015;**112**(31):9662-9667

[40] Johnson, C. K., Hitchens, P.L.,Pandit, P. S., Rushmore, J., Evans, T., Smiley, Y., Cristine, C. W., Doyle, M. M.(2020). Global shifts in mammalian population trends reveal key predictors of virus spillover risk. Proceedings of the Royal Society B2872019273

[41] Kreuder JC, Hitchens PL, Smiley ET, Goldstein T, Thomas K, Clements A, et al. Spillover and pandemic properties of zoonotic viruses with high host plasticity. Scientific Reports. 2015;5:14830 https://doi.org/10.1038/srep14830

[42] Wolfe ND, Daszak P, Kilpatrick AM, Burke DS. 2005 Bushmeat, hunting, deforestation, and prediction of zoonotic disease emergence. Emerging Infectious Diseases11, 1822-1827. (doi:10.3201/eid1112.040789)

[43] Brown, C. C. and Slenning, B. D. (1996). Impact and risk of foreign animal diseases. Journal of the American Veterinary Medical Association, 208(7): 1038-1040.

[44] Thomson G, Vosloo W, andBastos A. Foot and mouth disease in wildlife. Virus Research. 2003;**91**(1):145-161

[45] Arzt J, Baxt B, Grubman M, Jackson T, Juleff N, Rhyan J, et al. The pathogenesis of foot-and-mouth disease II: Viral pathways in swine, small ruminants, and wildlife; Myotropism, chronic syndromes, and molecular virus–host interactions. Transboundary and Emerging Diseases. 2011;**58**(4):305-326

[46] Dawe P, Flanagan F, Madekurozwa R, Sorensen K, Anderson E, Foggin C, et al. Natural transmission of foot-and-mouth disease virus from African buffalo (Syncerus caffer) to cattle in a wildlife area of

Zimbabwe. The Veterinary Record. 1994;**134**(10):230-232

[47] Bronsvoort B, Parida S, Handel I, McFarland S, Fleming L, Hamblin P, et al. Serological survey for foot-and-mouth disease virus in wildlife in eastern Africa and estimation of test parameters of a nonstructural protein enzyme-linked immunosorbent assay for buffalo. Clinical and Vaccine Immunology. 2008;**15**(6):1003-1011

[48] Vosloo W, Thompson P, Botha B, Bengis R, andThomson G. Longitudinal study to investigate the role of impala (*Aepyceros melampus*) in foot-and-mouth disease maintenance in the Kruger National Park, South Africa. Transboundary and Emerging Diseases. 2009;**56**(1-2):18-30

[49] Atuman YJ, Kudi CA, Abdu PA, Okubanjo OO, Abubakar A, Wungak Y, et al. Seroprevalence of foot and mouth disease virus infection in some wildlife and cattle in Bauchi state, Nigeria. Veterinary medicine international. 2020;*2020*:3642793 https://doi.org/10.1155/2020/3642793

[50] Dyason E (2010). Summary of foot-and-mouth disease outbreaks reported in and around the Kruger National Park, South Africa, between 1970 and 2006. *Journal of the South African VeterinaryAssociation*81(4):201-206(En.).

[51] Keet D, Michel A, Meltzer DGA. Tuberculosis in free-ranging lions (Panthera leo) in the Kruger National Park. Proceedings of the South African Veterinary Association Biennial Congress. 2000:232-241

[52] Keet DF, Hunter P, Bengis RG, Bastos AD, Thomson GR. The 1992 foot-and-mouth disease epizootic in the KrugerNational Park. Journal of the South African Veterinary Association. 1996;**67**:83-87

[53] Wungak, Y.S., Olugasa, B.O., Ishola, O.O., Lazarus, D.D. and Ularamu,, G.H "Foot-and-mouth disease (FMD) prevalence and exposure factors associated with seropositivity of cattle in north-central, Nigeria," African Journal of Biotechnology, vol. 15, no. 24, pp. 1224-1232, 2016.

[54] Madege, M.J. (2018). Current status of foot and mouth disease in Tanzania. rr-africa.oie.int › wp-content › uploads › 2018/07 › 19...Accessed on 14 September, 2020.

[55] Jori F, Alexander KA, Mokopasetso M, Munstermann S, Moagabo K, andPaweska JT. Serological evidence of Rift Valley fever virus circulation in domestic cattle and African buffalo in northern Botswana (2010-2011). Frontiers in Veterinary Science. 2015;**2**:63-71

[56] Ellis DS, Shirodaria P, Fleming E, Simpson DI. Morphology and development of Rift Valley fever virus in Vero cell cultures. Journal of Medical Virology. 1988;**24**:161-174

[57] Moiane B, Mapaco L, Thompson P, Berg M, Albihn A, Fafetine J. High seroprevalence of Rift Valley fever phlebovirus in domestic ruminants and African Buffaloes in Mozambique shows need for intensified surveillance. Infection Ecology & Epidemiology. 2017;**7**(1):1416248

[58] Opayele AV, Odaibo GN, Olaleye OD. Rift valley fever virus infection among livestock handlers in Ibadan, Nigeria. Journal of Immunoassay and Immunochemistry. 2018;**39**(6):609-621

[59] Van den Bergh, C., Venter, E.H., Swanepoel, R. and Thompson, P.N (2019). High seroconversion rate to Rift Valley fever virus in cattle and goats in far northern Kwa Zulu-Natal, South Africa, in the absence of reported

outbreaks. PLoS Neglected Tropical Diseases 13(5): e0007296.https://doi.org/10.1371/journal. pntd.0007296

[60] Food and Agriculture Organization (FAO) (2010). Rift Valley fever: transmission and spread. Causes of transmission. *FAO*, Rome (www.fao.org/doc.rep/006/y4611e/y4611e04).

[61] Bishop D, Calisher C, Casals J, Chumakov M, Gaidamovich SY, Hannoun C, et al. *Bunyaviridae*. Intervirology. 1980;**14**:125-143

[62] Olive M-M, Goodman SM, andReynes J-M. The role of wild mammals in the maintenance of Rift Valley fever virus. Journal of Wildlife Diseases. 2012;**48**(2):241-266

[63] Britch SC, Binepal YS, Ruder MG, Kariithi HM, Linthicum KJ, Anyamba A, et al. Rift Valley fever risk map model and seroprevalence in selected wild ungulates and camels from Kenya. PLoS One. 2013;**8**(6):e66626

[64] Paweska JT. Rift Valley Fever *Review ScientificTechnique Office International des Epizootics*. 2015;**34**(2):375-389

[65] Bolajoko M. and Atuman, Y. J (2019). Rabies ecology in Bauchi state, Nigeria Journal of Veterinary and Biomedical Sciences Vol. 2 (1). PP 5- 12, December, ISSN:2635-3067

[66] Real, L.A. and Childs. J.E., (2005). Spatial-temporal dynamics of rabies in ecological communities. Coll., 12, 170-186

[67] Smith JS, Sumner JW, Roumillat LF, Baer GM, Winkler WG. Antigenic characteristics of isolates associated with a new epizootic of raccoon rabies in the *U.S.* The Journal of Infectious Diseases. 1984;**149**:769-740

[68] Ballard WB, Follmann EH, Ritter DG, Robards MD, Cronin MA. Rabies and canine distemper in an arctic fox population in Alaska. Journal of Wildlife Diseases. 2001;**37**:133-137

[69] Sihvonen L. Documenting freedom from rabies and minimizing the risk of rabies being re-introduced to Finland. Rabies Bulletin Europe. 2003;**27**:5-6

[70] Ruppprecht CE, Willoughby R, Slate D. Current and future trends in the prevention, treatment and control of rabies. Expert Rev. Anti-Infect. Therapy. 2006;**4**(6):1021-1036

[71] Isoun TT, Losos GJ, Ikede BO. Diseases of zoo animals in Nigeria. Journal of Wildlife Diseases. 1972;**8**:335-339

[72] Okoh, A.E.J. (1976). Cases of rabies in captive caracal lynz (felis caracal) in Jos, Nigeria. Niger. Veter. J., 5, 52-56

[73] Kisali OB. Rabies in a civet cat (Virtacirtta); case report. Veter. Rec. 1977;**100**:291

[74] Enurah, Leonard & Ocholi, Reuben & Adeniyi, K.O. &Ekwonu, M.C.. (1988). Rabies in a civet cat (Civettictiscivetta) in the Jos Zoo, Nigeria. The British veterinary journal, 144. 515-6. 10.1016/0007-1935(88)90092-9.

[75] Nel LH, Sabeta CT, von Teichman B, Jaftha JB, Rupprecht CE, Bingham J. Mongoose rabies in southern Africa, a re-evaluation based on molecular epidemiology. Virus Research. 2005;**109**:165-173

[76] Dzikwi AA, Kuzmin II, Umoh JU, Kwaga JKP, Ahmad AA, Rupprecht CE. Evidence of Lagos bat virus circulation among fruit bats. J. wild life Dis. 2010;**46**(1):267-271

[77] Luis, A.D., Hayman D.T.S. and Webb C.T. (2013). A comparison of bats and rodents as reservoirs of

zoonotic viruses: Are bats special? Proc. Roy. Soc. Biol. Sci., 280 (1756): 20122753doi:10.1098/rspb.2012.2753

[78] Atuman, Y.J., Adawa, D.A.Y., Okewole, P.A., Shamaki, D., Audu, S.W., Mshelbwala, P. and Ogunkoya, A.B. (2014). Detection of rabies antigens in the brain tissues of jackals and mongooses and its implications on public health and conservation goals in Bauchi state, Nigeria A R T I C L E I N F O. 3. 42-47. 10.14196/sjvs.v3i3.12 (B)

[79] Zulu GC, Sabeta CT, Nel LH. Molecular epidemiology of rabies, focus on domestic dogs (Canis familiaris) and black-backed jackals (*Canis mesomelas*) from northern South Africa. Virus Research. 2009;**140**(1-2):71-78

[80] Stuchin, M., Machalaba, C., Olival, K., Artois, M., Bengis, R., Diaz, F., Erlacher V. E., Forcella, S., Leighton, F., Murata, K., Karesh, W. and Tizzani, Paolo. (2018). Rabies as a threat to wildlife. *Revue scientifique et technique (International Office of Epizootics)*. 37. 241-257. 10.20506/rst.37.2.2858.

[81] Walsh MG, Mor SM, Hossain S. The elephant-livestock interface modulates anthrax suitability in India. Proceedings. Biological sciences. 2019;*286*(1898):20190179 https://doi.org/10.1098/rspb.2019.0179

[82] O'Reilly, L.M and Daborn, C.J (1995). The epidemiology of *M. bovis* infections in animals and man: A review. Tubercle and Lung Diseases, 76 (Supplement 1): 1-46.Veter. J., 5, 52-56

[83] Etter E, Donado P, Jori F, Caron A, Goutard F, Roger F. Risk analysis and bovine tuberculosis,a re-emerging zoonosis. Annals of New York Academy of Science. 2006;**108**(1):112

[84] Grange JM. Mycobacterium bovis infection in human beings. Tuberculosis. 2001;**81**(1-2):71-77

[85] Brosch R, Gordon SV, Marmiesse M. A new evolutionary scenario for the mycobacterium tuberculosis complex. Proceedings of the National Academy of Science of the United State of America. 2002;**99**(6):3684-3689

[86] Cosivi, O., Meslin, F.X, Daborn, C.J., Grange J.M (1995). Epidemiology of *Mycobacterium bovis* infection in animals and humans, with particular reference to Africa. Rev Sci Tech, 14:733-46

[87] Erwin PC, Bemis DA, Mawby DI, McCombs SB, Sheeler LL, Himelright IM, et al. Mycobacterium tuberculosis transmission from human to canine. Emerging Infectious Diseases. 2004;*10*(12):2258-2260 https://dx.doi.org/10.3201/eid1012.040094

[88] de Lisle GW, Bengis RG, Schmitt SM, O'Brian DJ. Tuberculosis in free-ranging wildlife: Detection, diagnosis and management. Revue Scientifique et Technique. 2002;**21**:317-334

[89] Atuman YJ, Ibrahim S, Abubakar AI, Kudi CA. Seroprevalence of bovine tuberculosis and brucellosis in Agro-pastoralist livestock herds and wildlife in Yankari game reserve: Public health implications. Sokoto Journal of Veterinary Sciences. 2018;**16**(2):83-90

[90] Nishi JS, Shury T, Elkin BT. Wildlife reservoirs for bovine tuberculosis (*Mycobacterium bovis*) in Canada: Strategies for management and research. Veterinary Microbiology. 2006;**112**(2-4):325-338

[91] Mfinanga SG, Morkre O, Kazwala RR. The role of livestock keeping in tuberculosis trend in Arusha, Tanzania. International Journal of Tuberculosis and Lung Disease. 2003;7(7):695704

[92] Munyeme M and Muanangandu (2011). A review of bovine tuberculosis

in the Kafue basin ecosystem. *Veterinary Medicine International*, doi:10.4061/2011/918743.

[93] Gomo, C.,Musari,. S., de Garine-Wichattsk., Caron, M., Pfukenyi, DM. and Van Heerden, H. Detection of Brucella abortus in Chiredzi district Zimbabwe. Onderstepoort Journal of Veterinary Research. 2012;**79**(1):417-422

[94] Katale BZ, Mbugi EV, Karimuribo ED, Keyyu JD, Kendall S, Kibiki GS, et al. Prevalence and riskfactors for infection of bovine tuberculosis in indigenous cattle in the serengeti ecosystem, Tanzania. BMC Veterinary Research. 2013;**9**(1):267

[95] Daszak P, Cunningham AA, Hyatt AD. Anthropogenic enviromental change and the emergence of infectious diseasein wildlife. Acta Tropica. 2001;**78**(2):103-116 https://dx.doi.org/10.3201/eid1012.040094

[96] Godfroid, J., Clocakacrt, A. and Liqutard J.P (2005). From the discovery of the Malta fever's agent to the discovery of a marine reservoiur, brucellosis has been continuous re-emerging zoonoses. Veterinary Research,Doi: 10.1051/vetres:2005003.

[97] Pappas G, Papadimitriou P, Akritidis N, Christou L, Tsianos EV. The new global map of human brucellosis. Lancet Infectious Diseases. 2006;**6**(2):91-99

[98] Maxwell JR, Bill DE. Developing a brucellosis public health information and awareness campaign in Iraq. Military Medicine. 2008;**17**(1):79-84

[99] Hugh-Jones, M. and Vos, V. (2002). Anthrax and Wildlife. Revue scientifique et technique (International Office of Epizootics). 21. 359-83. 10.20506/rst.21.2.1336.

[100] Van Wyk IC, Boomker J. Parasites of south African wildlife: XIX. The prevalence of helminths in some common antelopes, warthogs and a bushpig in the Limpopo province, South Africa. Ondersteoort Journal of Veterinary Research. 2011;**78**(1):1-11

[101] Wall DT. Anaplasmosis control and diagnosis in South Africa. Annals of the New York Academy of Sciences. In: 10.1111/j.17496632.2000.tb05327. 2000

[102] Ogunji FD, Akinboade OA, Dipeolou OO, Ayeni J, andOkaeme AN. The prevalence of gastrointestinal parasites and bacteria in the game scouts at the Kainji Lake National Park of Nigeria. International Journal of Zoonoses. 1984;**11**(1):119-122

[103] Muriuki SMK, Murugu RK, Munene E, Karere GM, Chai DC. Some gastro-intestinal parasites of zoonotic (public health) importance commonly observed in old world non-human primates in Kenya. Acta Tropica. 1998. DOI: 10.1016/s0001-706x(98)00040-0

[104] Oyeleke SB, Edungbola OJ. Prevalence of gastrointestinal helminths of wild animals in Kainji lake National Park and federal college of wildlife management new Bursa,Niger state, Nigeria. Nigerian Journal of Parasitology. 2001;**22**(2):129-136

[105] Karere G, Munene E. Some gastro-intestinal tract parasites in wild De Brazza's monkeys (Cercopithecus neglectus) in Kenya. Veterinary Parasitology. 2002;**110**(1):153-157

[106] Moudgil AD, Singla LD. Role of neglected wildlife disease ecology in emergence and resurgence of parasitic diseases. Trends in Parasitology Research. 2013;**2**(2):18-23

[107] Kalema-Zikusoka G, Kock R, andMacfie E. Impenetrable National Park, Uganda. The Veterinary Record. 2002;**150**:12-15

[108] Jongejan F, Uilenberg G. The global importance of ticks. Parasitology. 2004;**129**:S3-S14

[109] Nizeyi JB, Innocent RB, Erume J, Kalema GR, Cranfield MR, andGraczyk TK. *Campylobacteriosis, Salmonellosis*, and *Shigellosis* in free-ranging human-habituated mountain gorillas of Uganda. Journal of Wildlife Diseases. 2001;**37**(2):239-244

[110] Mutani A, Rhynd K, andBrown G. A preliminary investigation on the gastrointestinal helminths of the Barbados green monkey, Cercopithecus aethiops sabaeus. Revista do Instituto de Medicina Tropical de São Paulo. 2003;**45**(4):193-195

[111] Munang'andu HM, Siamudaala VM, Munyeme M, Nalubamba KS. Detection of parasites and parasitic infections of free-ranging wildlife on a game ranch in Zambia: A challenge for disease control. Journal of Parasitology Research. 2012;**9**:1-12

[112] Malan F, Horak I, De Vos V, andVan Wyk J. Wildlife parasites: Lessons for paasite control in livestock. Veterinary Parasitology. 1997;**71**(2):137-153

[113] Conradie, I. (2008). The prevalence of helminths in warthogs, bushpigs and some antelope species in Limpopo province, South Africa. University of Pretoria. pp 113-119

[114] Agbede RI. A survey of ectoparasites and ectoparasitic condition of animals in Zaria. Nigerian Journal of Animal Production Resources. 1981;**1**(2):179-180

[115] Gwakisa P, Yoshihara K, Long TT, Gotoh H, Amano F, Eiichi M. Salivary gland extract of *Rhipicephalus appendiculatus* ticks inhibits in vitro transcription and secretion of cytokines and production of nitric oxide by LPS stimulated JA-4 cells. Veterinary Parasitology. 2001;**99**:53-61

[116] Biswas, S (2003). Role of veterinarians in the care and management during harvest of skin in livestock species. In: Proceedings National Seminar on Leather Industry in Today's Perspective, Kolkata, India, pp. 62-64.

[117] de Castro JJ. Sustainable tick and tick-borne disease control in livestock improvement in developing countries. Veterinary Parasitology. 1997;**71**:77-97

[118] Odunlami, S. S. andLake, B. (2003). An assessment of the ecotourism potential of Yankari national park, Nigeria. Eco Club. Communication E paper Series NR, 7.

[119] Messenger AM, Barnes AN, andGray GC. Reverse zoonotic disease transmission (zooanthroponosis): A systematic review of seldom-documented human biological threats to animals. PLoS One. 2014;**9**(2):e89055

[120] VanderWaal K, Omondi GP, andObanda V. Mixed-host aggregations and helminth parasite sharing in an east African wildlife–livestock system. Veterinary Parasitology. 2014;**205**(1):224-232

[121] Rhyan J, andSpraker T. Emergence of diseases from wildlife reservoirs. Veterinary Pathology. 2010;**47**(1):34-39

[122] Thompson R, Lymbery A, andSmith A. Parasites, emerging disease and wildlife conservation. International Journal for Parasitology. 2010;**40**(10):1163-1170

[123] Ghandour AM, Zahid NZ, Banaja AA, Kamal KB, Bouq AI. Zoonotic intestinal parasites of hamadryas baboons Papio hamadryas in the western and northern regions of Saudi Arabia. Journal of

Tropical Medicine and Hygiene. 1995;**98**(6):431-439

[124] Murray S, Stem C, Boudreau B, Goodall J. Intestinal parasites of baboons (Papio cynocephalus anubis) and chimpanzees (pan troglodytes) in Gombe National Park. Journal of Zoo and Wildlife Medicine. 2000;**31**(2):176-178

[125] Alvarado-Rybak M, Solano-Gallego L, andMillán J. A review of piroplasmid infections in wild carnivores worldwide: Importance for domestic animal health and wildlife conservation. Parasites and Vectors. 2016;**9**(1):538

[126] Eneji MA, Ode FA, Bullus ML. Diversification of Nigeria's economy; impact of tourism on sustainable development in Nigeria. International Journal of Research in Humanities and the Social Studies. 2016;**3**(5):36-44

[127] Shock BC, Murphy SM, Patton LL, Shock PM, Olfenbuttel C, Beringer J, et al. Distribution and prevalence of Cytauxzoonfelis in bobcats (*Lynx rufus*), the natural reservoir, and other wild felids in thirteen states. Veterinary Parasitology. 2011;**175**(3):325-330

[128] Najm N-A, Meyer-Kayser E, Hoffmann L, Herb I, Fensterer V, Pfister K, et al. A molecular survey of Babesia spp. and Theileria spp. in red foxes (Vulpes vulpes) and their ticks from Thuringia, Germany. Ticks and Tick-borne Diseases. 2014;**5**(4):386-391

[129] Duscher GG, Fuehrer HP, andKübber-Heiss A. Fox on the run– molecular surveillance of fox blood and tissue for the occurrence of tick-borne pathogens in Austria. Parasites and Vectors. 2014;**7**(1):521-531

[130] Hofshagen M, Aavitsland P, Kruse H. Trends and Sources of Zoonotic Agents in Animals, Feedingstuffs, Food, and Man in Norway, Report to the EU. Oslo: Norwegian Department of Agriculture; 2003

[131] Henttonen H, Fuglei E, Gower CN, Haukisalmi V, Ims RA, Niemimaa J. *Echinococcus multilocularis* on Svalbard: Introduction of an intermediate host has enabled the local life-cycle. Parasitology. 2001;**123**:547-552

[132] Melton DA, Melton CL. Condition and mortality of waterbuck (Kobus illipsiprymus) in the Umfolozi game reserve. African Journal of Ecology. 1982;**20**(1):89-103

[133] Muruthi CW, Lwande OW, Makumi JN, Runo S, Otiende M. Phenotypic and genotypic identification of ticks sampled from wildlife species in selected conservation sites of Kenya. Journal of Veterinary Science and Technology. 2016;**7**:293

[134] Sumbria D, Singla LD. Thwack of worldwide weather transformation on vector and vector-borne parasitic infections. ARC J. Anim. Vet. Sci. 2017;**3**(2):1-10

Chapter 5

Interlinks between Wildlife and Domestic Cycles of *Echinococcus* spp. in Kenya

Dorothy Kagendo, Eric Muchiri, Peter Gitonga and Esther Muthoni

Abstract

Effective conservation and management of wildlife in the current changing world, call for incorporation of infectious zoonotic diseases surveillance systems, among other interventions. One of such diseases is echinococcosis, a zoonotic disease caused by Echinococcus species. This disease exists in two distinct life cycle patterns, the domestic and wildlife cycles. To investigate possible inter-links between these cycles in Kenya, 729 fecal samples from wild carnivores and 406 from domestic dogs (*Canis lupus familiaris*) collected from Maasai Mara and Samburu National Reserves were analyzed. Taeniid eggs were isolated by zinc chloride sieving-flotation method and subjected to polymerase chain reaction of nicotinamide adenine dehydrogenase subunit 1 (NAD1). Subsequent amplicons were sequenced, edited and analyzed with GENtle VI.94 program. The samples were further subjected to molecular identification of specific host species origin. All sequences obtained were compared with those in Gene-bank using Basic Local Alignment Search Tool (BLAST). The study found that there were 74 taeniid positive samples, 53 from wild carnivores and 21 from domestic dogs. In wildlife, mixed infections with *Echinococcus* and *Taenia* species were identified and these included *E. granulosus sensu stricto*, *E. felidis, T. canadensis* G6/7, *Taenia hydatigena*, *T. multiceps*, and *T. saginata.* Domestic dogs harbored *Echinococcus and Taenia* species similar to wild carnivores including *E. granulosus* G1–3, *E. felidis, T. multiceps, T. hydatigena*, and *T. madoquae*. *Taenia species* of nine taeniid eggs were not identified. Majority of genotypes were found in hyena (*Crocuta crocuta*) fecal samples. Distribution of *Echinococcus* and *Taenia* spp. varied with hosts. Mixed infections of *Echinococcus spp*, *T. multiceps* and *T. hydatigena* in a single animal were common. There seemed to be existence of interactions between the two cycles, although public health consequences are unknown. The presence of *T. saginata* in hyena suggests scavenging of human fecal matter by the animal. In addition, presence of *T. multiceps, T hydatigena, T madoquae* and *T. saginata* in the two cycles suggested possible human exposure to these parasites. The results are important in drawing up of strategies and policies towards prevention and control of *Echinococcosis* and other *Taenia* related parasitic infections, especially in endemic areas given their potential risk to public and socio- economic livelihood.

Keywords: conservation, management wildlife, Echinococcosis, Taenia species

1. Introduction

1.1 Importance of wildlife conservation and management

Wildlife conservation and management is the process of caring for wild animal species and their environments from destruction, including preserving rare species from extinction. All this is done to sustain a better balance within an ecosystem as well as maintaining the beauty of mother nature [1–3]. In cases where the balance is interrupted, for example in communities where wild carnivores are killed due to wildlife-human conflicts, it may lead to overpopulation of wild-herbivores, and consequently translate into overgrazing of the available vegetation and deforestation [4, 5]. For centuries wildlife, has been reported to serve as a source of food, thus sustaining human life through provision of products such as honey and bush meat [6]. Where strict wildlife management procedures are observed, the chances for transmission of zoonotic diseases are reduced and therefore, good health and disease-free populations [7]. Due to improved wildlife conservation and management strategies, the economy of many countries globally has improved due to income generated from tourism attraction [3]. Tourist visits have in turn led to enhanced social and cultural livelihood in different communities, the Maasai and Samburu of Kenya included [3, 8].

2. Challenges facing wildlife species

In Africa wildlife has faced great challenges, often attributed to human activities including encroachment into wildlife sanctuaries and loss of habitats [9]. Other challenges include poaching and illegal wildlife trade, activities that have led to the declining numbers of wild animals' overtime [8, 10, 11]. Besides loss of habitats, poaching, pollution, climate change and invasive species, emerging and re-emerging zoonotic diseases are increasingly featuring as a major challenge in wildlife conservation. The current wild pandemic of covid-19 is a major example of such diseases, transferable from animals to human or/and vice versa which often happen when human encroach wildlife sanctuaries, and affect the balance of Nature, for example, deforestation and modification of natural habitats as a result of land use and land cover changes is responsible for outbreak of about 50% of the emerging zoonoses [12].

3. Emergence of infectious zoonotic diseases

Wide range of pollutants affects wildlife health and sometimes lead to animal death. Diseases in wildlife influence several biological factors like reproduction, survival fitness and abundance of wildlife species [13]. Often arthropods and other animal species of wildlife origin have been reported to transmit diseases including Nile fever, Lyme disease, Encephalomyelopathies, COVID-19, Bovine tuberculosis, among other zoonotic diseases. Ben (2014) stated need for humans to refrain from anthropocentric attitudes towards wildlife and learn a need for respect to ecosystems, emphasizing on major benefits that exist when the balance in nature is maintained. In their report, Vila and group of scientists reported endoparasites causing zoonotic diseases in cattle and wild animals in Europe [14, 15]. The Asian tiger mosquito (*Aedes albopictus)* was reported as a vector that caused over 22 Arboviruses worldwide. The mosquito has been reported to have caused outbreaks of dengue and chikungunya in Northern Italy [16]. During the time, the dengue fever was cited as a major cause of deaths in children in moat of the Asian countries [16]. In the African continent, tsetse fly (*genus Glossina*) has been reported to cause trypanosomiasis in both humans and livestock [15, 17, 18]. Simwango et al. [18] linked exposure of the Maasai people to zoonotic diseases, with their frequent interactions with wildlife. A recent emerging zoonotic disease, COVID-19, caused by Corona virus with impacts to over 210 countries worldwide [12, 19], is linked to animal human transmission cycle [20, 21]. The current emergence of viruses, parasites and bacteria as significant pathogens, originate mainly from human encroachment areas [22]. These organisms had the capability of reducing body immunity and causing acute illnesses that could often be fatal. Helminths, trematodes and cestodes are important parasitic human-wildlife diseases. In East Africa most of the diseases are augmented by the closeness of pastoralists with their livestock into wildlife sanctuaries, especially during cattle herding [23]. However, only limited data on interlinks between human and wildlife disease cycles exist. The impact of emerging and re-emerging zoonotic diseases is a nightmare, which continues to cause heavy pandemics worldwide, more effect being felt in developing countries including Kenya [24]. It is worth noting that zoonotic diseases found in human-wildlife interfaces are complex, and thus hard to predict on time.

4. Echinococcosis: a zoonotic diseases

Cystic echinococcosis (CE) is a zoonotic disease of human and animals (livestock and wildlife), caused by larval stages of tapeworms of dogs and other carnivores. The disease occurs worldwide, but is particularly prevalent under conditions of extensive livestock keeping, uncontrolled slaughter and low levels of hygiene [25]. In sub-Saharan Africa, CE is a serious public health and economic problem in the eastern and southern parts, especially for pastoralists and nomadic communities, but reliable data are limited [25]. Effective control is prevented by inadequate resources and limited knowledge about the epidemiology. Several Hydatid cysts may occupy space on a lung, liver or kidney making it difficult for the person or animal to breath. The parasite exists in two distinct life cycle patterns, namely the domestic and the wildlife cycles [9].

Humans get cystic echinococcosis after ingestion of Taeniid eggs that may have been shed through feces of domestic dogs (in the domestic cycle) and/or wild carnivores in the wildlife cycle. *Echinococcus granulosus* s.l. is a cestode parasite of the family *Taeniidae.* The parasite is made up of at least five species, namely; *E. granulosus* sensu stricto, *E. equinus*, *E. ortleppi*, *E. canadensis* and *E. felidis.* Distribution of these cestode taxa vary greatly across the globe. However sub-Saharan Africa is by far the most diverse region with all species of *E. granulosus* sensu lato found, with exception of the genotypes G8 and G10 of *E. canadensis.* [26, 27].

Globally. *granulosus* sensu stricto (s. s). is the most important agent for human CE in both humans and animals [28, 29]. Contributions of *E. equinus* and *E. felidis* in human CE is non-existent, that of *E. ortleppi* is very marginal and *E. canadensis* G6/7 is only about 11% [27, 29]. The case report of genotype *E. granulosus* Gomo from an Ethiopian patient by Wassermann *et al.,* reported in 2016 as well as the prevalence of several *E. granulosus* taxa in countries such as Kenya present aspects of the disease that is not yet fully understood.

In Kenya, it has been unveiled that the two transmission patterns of *Echinococcus* exist and an initial observation of an interface was reported previously [9]. Global control of CE in domestic settings is very complex and presents a variation of challenging factors in endemic regions such as illiteracy, poor road networks, social cultural beliefs, and poverty [22, 30]. In parts of Africa, control of CE has only been partially achieved despite establishment of long-term control programs [31, 32]. The diversity of species, a wide range of hosts and various cultural practices in sub-Saharan Africa have made control strategies of CE in the region less successful. Therefore, sylvatic-domestic transmission interface presents a new aspect of Echinococcus species that is the least understood. In Africa, where diversity of *E. granulosus* (*sensu lato*) is very high, elucidation of the sylvatic-domestic interaction is very essential. A recent study reported *E. felidis*, a strain well adapted to lions in the wildlife and also a sister species to the global problematic *E. granulosus* (sensu stricto) in domestic dogs [29]. The pathogenicity of *E. felidis* to domestic animals remains unknown. In 2014, Kagendo et al., isolated *E. granulosus* s. s. eggs from lion feces, however the extent of actual transmission in the wildness or how the lions contracted the taxa was a mare speculation, since there were only a few reports showing the taxa to have been isolated in a stool sample from a warthog [33]. The present study aimed to evaluate the interaction of the sylvatic and domestic cycles of this zoonotic disease in areas adjacent to the national reserves in Kenya.

5. Materials and methods

5.1 Study areas

The study was done in two cystic Echinococcosis (CE) endemic areas of Maasai Mara and Samburu National reserves. The Maasai Mara National Reserve, situated in the northern part of Tanzania's Serengeti National Park occupies 1500 km^2 [9]. The Reserves a part of the Greater Serengeti-Mara Ecosystem which is globally popular for unique phenomenon of wildebeest's migration. The ecosystem has suitable vegetation and climatic conditions supporting a variety of wild animals, livestock and human beings. In this case, co-existence of wild animals with pastoral communities in the area is evident [9].

Samburu national reserve covers about 165 km^2. Human beings, livestock and wild animals in are primarily dependent on the river 'Ewaso Nyiro'. Human and wildlife interactions are therefore a common phenomenon, with wild carnivores often preying on livestock and humans fighting back and killing the predators.

5.2 Collection of study samples and isolation of Taeniid eggs

Fecal samples of wild carnivores were collected from the environment by following signs and tracks [34]. Similarly, freshly dropped fecal samples of domestic dogs were collected within the homesteads in the two areas. Taeniid eggs were isolated from 3 g of the fecal samples using the Zinc floatation method and subsequent microscopy identification [35] (**Figures 1** and **2**).

5.3 Sample processing

5.4 DNA isolation for PCR

Individual taeniid eggs were picked under the microscope, lysed in 10 μl of 0.02 N NaOH solution. Lysates were used for amplification of the short fragment of NADH dehydrogenase Sub unit 1 gene (*nad1*) of *Echinococcus spp* and other *Taenia species* [36] (**Figure 3**).

5.5 Polymerase chain reaction and gene sequencing of *nad1* positive amplicons

Amplification of a 200 bp long fragment of *nad1* was done in a primary PCR using Nadnest A 5'-TGTTTTTGAGATCAGTTCGGTGTG3' and Nadnest C 5'

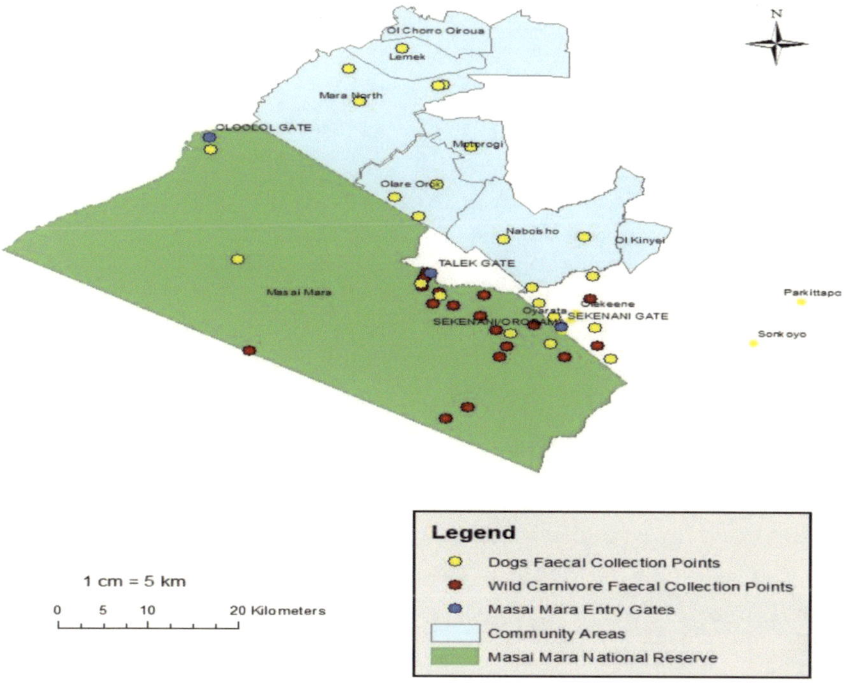

Figure 1.
The Maasai Mara wildlife human interface areas where samples were collected.

CATAATCAAACGGAGTACGATAG −3′ primers in a 25 μl mix that was constituted using 1x dream Taq green buffer (20 nM Tris–HCl pH 8), 0.2 mM dNTPs, 0.25 μM of each forward and reverse primer, 2 mM $MgCl_2$ and 0.625 U of dream Taq green DNA polymerase (Thermo scientific) and 2 μl of the target DNA template. A nested PCR was done using Nadnest 5'-B CAGTTCGGTGTGCTTTTGGGTCTG-3′ and Nadnest D 5'-GAGTACGATTAGTCTCACACAGCA primers in a 25 μ mixture of same constitution as the primary PCR the use of 1 μl of the of the primary PCR amplicon as a source of DNA template [33]. Cycling conditions of primary and nested PCRs were the same; initial denaturation for 5 minutes at 94°C, and a 35-cycle involving denaturation at 94 ^{0}Cfor 30 s, elongation at 55°C for 30 s and annealing at 72°C for 1 minutes., and a final extension at 72°C for 5 minutes. Detection of amplicons was done on a 2% gel red stained agarose gel. All *nad1*positive individual samples were purified using high pure product purification kit (Roche, Germany) and sequenced using the reverse primer at GATC Biotech AG, Germany.

5.6 DNA sequence analysis and taeniid parasite identification

DNA Sequences were viewed and edited using the GENtle software (Manske M. 2003, University of Cologne, Germany). Clean DNA sequences were then compared with existing sequences in the NCBI GenBank using the Basic Local Alignment Search Tool (BLAST).

5.7 Wild carnivore host identification

Host specificity of all taeniid positive samples from the environment of the parks were done by a method previously described [33]. A PCR system using

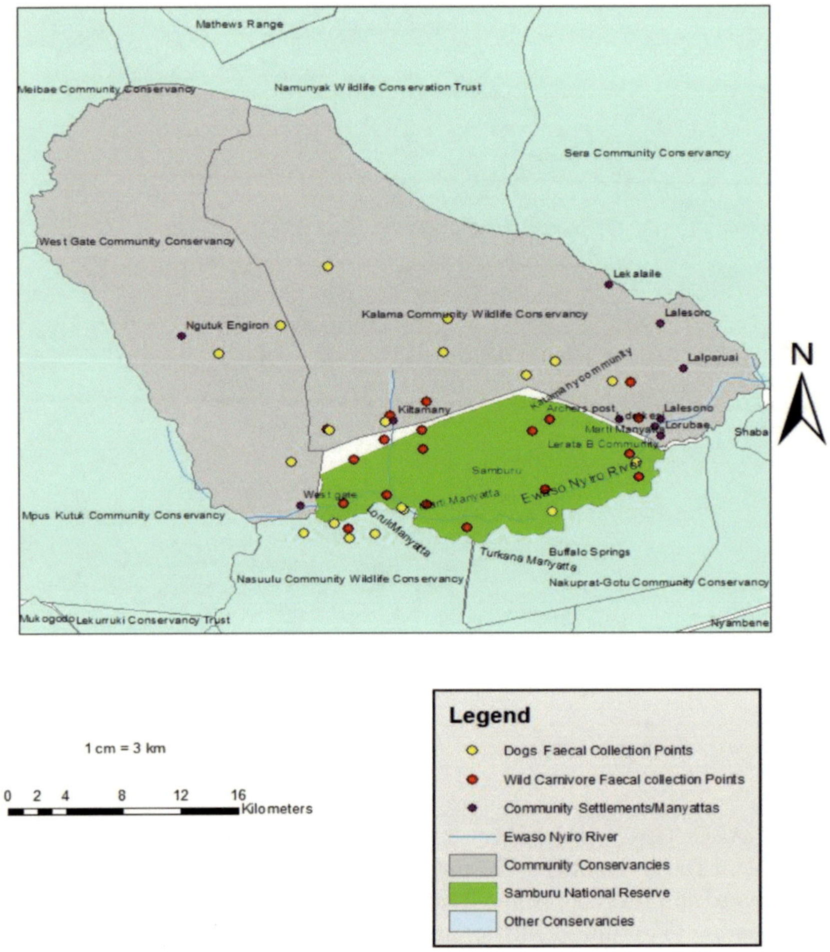

Figure 2.
The Samburu wildlife human interface areas where samples were collected.

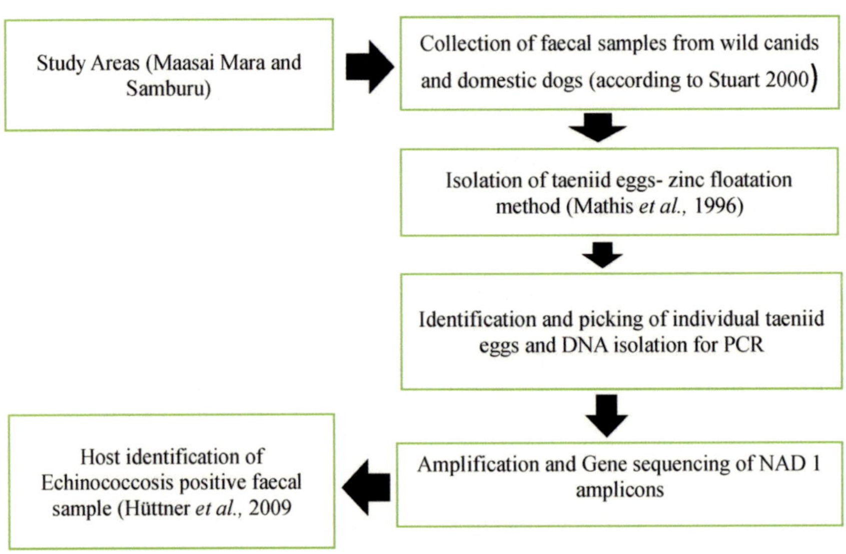

Figure 3.
Steps during fecal sample processing.

primer pairs forward 5'-TCATTCATTGA(C/T) CT(C/T) CCCAC(C/T) CCA-3'and reverse 5'-ACGGTA(A/G) GACATA(A/T) CC(C/T) ATGAA(G/T) G-3′ for primary reaction and a secondary reaction with primer pairs forward CA(C/T) CCAA(C/T) ATCTCAGCATGAA and reverse 5′-(G/T) GC(G/T) GTAGCTAT(A/T) ACTGTGAA(C/T) A(A/G)-3′ were used to amplify partial fragment of the *cob* gene. A different primer pair was used for amplification of cob sequence of domestic dogs including lupus for *cob* 5'-CATCTAACATCTCTGCTTGATG-3'and lupus rev 5'-CTGTGGCTATGGTTGCGAATAA-3′. The subsequent cob PCR amplicons were purified and sequenced, then used in identification of host origin by comparing to earlier gene bank entries including; hyena (NC_020670), leopard (NC_010641), lion (KC495058) and domestic dogs (NC 002008).

6. Results

6.1 *Echinococcus* and *Taenia* spp. in wild carnivores of Maasai Mara and Samburu national reserves

A total of 729 fecal samples of wild carnivores from Maasai Mara (387) and Samburu (342) were screened for *taeniid* eggs and subsequently characterized to the cestode species level. Of these 53 fecal samples contained *taeniid* eggs, out of which 521 eggs were isolated. Each egg was treated as isolate in the subsequent molecular analysis. All isolated eggs were screened by a PCR test for *Taeniidae* amplification of a partial fragment of the NADH dehydrogenase subunit 1 (*nad1*) which yielded 183/521 (35%) taeniid positive from the two parks; Maasai Mara (86/183) and Samburu (97/183).

DNA sequence analysis of the t*aeniid* eggs revealed occurrence of *E. granulosus* (G1-G3) and *E. felidis* in Maasai Mara National Reserve. In Samburu National Reserves there were *E. granulosus* (G1-G3), *E. felidis,* and *E. canadensis* G6/7 (**Table 1**). Three *Taenia* spp. were identified in the two National Reserves -*Taenia multiceps,* and *T. hydatigena* from Maasai Mara and *T. hydatigena*, *T. multiceps* and *T. saginata* in Samburu (**Table 1**).

6.2 Confirmation of wild carnivore hosts origin of *Taeniid* positive samples

In addition to signs and tracks used in identifying the source of fecal samples in the field, the actual host origin of the 53 taeniid positive samples (26 from Maasai Mara and 27 from Samburu) were confirmed by PCR and DNA sequencing of the cob gene. The cob DNA sequences indicated the involvement of *Crocuta crocuta* (30/53), *Panthera leo* (7/53), *Canis lupus familiaris* (7/53), and *Canis adustus* (1/53) in the two National Reserves (**Table 1**). Host origin of 8 *taeniid* positive samples could not be determined.

6.3 *Echinococcosis* and *Taenia* spp. in the domestic settings in areas around Maasai Mara and Samburu national reserves

In the vicinity of Samburu National Reserve, 406 fecal samples from domestic dogs were collected; from 21 samples, 304 taeniid eggs were isolated. Ninety-two of the 304 eggs were positive on *nad1* PCR and revealed *E. granulosus* (G1-G3) (9) and *E. felidis* (47), T. *hydatigena* (10), *T. madoquae* (10), *T. multiceps* (7), and undetermined *Taenia* spp. (9). The domestic dog origin of all *E. felidis* positive fecal samples were confirmed by PCR and DNA sequencing. An earlier report where 500 domestic dog fecal samples from Maasai Mara National Reserve were screened, 34 samples

Park	Animal host	n taeniid positive / n samples	n taeniid positive PCR / n eggs screened	*Echinococcus* and *Taenia* spp.
Samburu	*Crocuta crocuta* (Spotted hyena)	13/342	51/156	30 *E. felidis*, 14 *E. granulosus* (G1-G3), 6 *T.hydatigena*, 1 *T. saginata*
	Panthera leo (Lion)	3/342	11/36	4 *E. felidis*, 5 *E. granulosus* (G1-G3), 2 *T. hydatigena*
	Canis lupus familiaris (Domestic dog)	6/342	13/72	8 *E. felidis*, 2 *E. granulosus* (G1-G3), 1 *E. canadensis* G6/7, 1 *T.hydatigena*, 1 *T. multiceps*
	Canis adustus (Side striped jackal)	1/342	5/12	2 *E. felidis*, 3 *E. granulosus* (G1-G3)
	Unidentified host	4/342	17/60	9 *E. felidis*, 7 *E. granulosus* (G1-G3), 1 *E. canadensis* G6/7
Maasai Mara	*Crocuta crocuta* (Spotted hyena)	17/387	61/197	41 *E. felidis*, 18 *T. hydatigena*, 2 *T. multiceps*
	Panthera leo (Lion)	4/387	12/48	9 *E. felidis*, 1 *T. hydatigena*, 2 *T. multiceps*
	Canis lupus familiaris (Domestic dog)	1/387	5/12	4 *E. felidis*, 1 *T. hydatigena*
	Unidentified host	4/387	8/48	3 *E. felidis*,1 *E. granulosus* (G1-G3), 2 *T. hydatigena*, 2 *T. multiceps*

Table 1.
Echinococcus *and* Taenia *spp. among carnivorous hosts of the Maasai Mara and Samburu National Reserves.*

were found positive for *nad 1*, of which 92/213 individual taeniid eggs were identified as *E. granulosus* (G1-G3) (86), *E. ortleppi* (2), *E. felidis* (3) and *E. canadensis* (1) [33].

7. Discussion

Human encroachment into wildlife sanctuaries has augment domestic-wildlife interactions thereby raising the risk margin for transmission of zoonotic diseases. Reduced interactions between human and wild animals by putting in place strict wildlife conservation and management legislations and strategic measures will reduce the burden of zoonotic disease transmission [37]. Population density in wildlife areas may occur in cases where management strategies include introduction of new animal species, which often result into introduction of new strains of zoonotic diseases, amidst improving number of animal population [38]. Wildlife movements often facilitate transfer of different disease strains from one point to the other, with an example of the wildebeest migration occurring every year from Serengeti to Maasai Mara [9]. While the migratory behavior is termed as a big economic gain due to increased tourist attraction for both countries, transmission of new strains that can cause extinction of wild species is possible. This is evidenced by a report in 2014 on existence of *Echinococcus granulosus* G1–3 in wildebeests [9]. The increased disease predisposition in wildlife sanctuaries has not only led to mortalities, and hence reduced wild animal populations but also diseases transmitted from wildlife to humans, example being the increasing burden of arboviruses and the current world pandemic of COVID-19 [12, 39, 40].

Zoonotic diseases have caused advanced effect especially in low and middle-income countries [41]. On average, up to 40% deaths occur in Africa due to infectious diseases, most of which are zoonotic [42]. These diseases have been reported to not only cause animal or human sickness but have led to deaths and major economic loses [37, 43]. Echinococcosis, a neglected zoonotic disease, has been reported to have highest prevalence in Kenya [30, 44]. It is hypothesized that the disease transmission could be minimized by improved wild conservation management systems in the country, since this has been seen to work well as reported in previous studies [43, 45]. Most wildlife sanctuaries are unfenced, and cattle are observed often at the heart of the protected areas, and wild animals in human homesteads [24]. This is equally interlinked with human bad slaughter behavior, where condemned offal is offered to domestic dogs, and with wild animals marauding at night, they may access and feed on this offal. This, consequently, leads to transmission of zoonotic diseases including Echinococcosis. The only reliable cure for Echinococcosis is a total removal of hydatid cyst, which is an extremely expensive undertaking, which calls for a specialized surgeon. Emergence and re-emergence of *Echinococcus* spp. has been reported in most countries in the African continent [24]. Until recently, six genotypes/species have been noted in Sub-Saharan Africa; *E. granulosus* sensu stricto, *E. canadensis* G6/7, *E. ortleppi*, *E. equinus*, *E. felidis* and *E. granulosus* genotype *G. omo* [9, 25–27, 33, 34]. These genotypes range from domestic origins (*E. granulosus* sensu stricto, *E. canadensis* G6/7, *E.ortleppi*, *E. granulosus* genotype Gomo), to wild origin spp. i.e.*felidis* and *E. equinus*. In this study, we confirmed the existence of the previously suggested overlap between domestic pets and wildlife cycles of *Echinococcus* species in Kenya.

During material sampling, it was observed that livestock herding inside the Maasai Mara National Reserve, took place at night as such the accompanying dogs might have access to carcasses of preyed animals. In both reserves the 'lion strain' *E. felidis* could be isolated from both wild carnivores and domestic dogs. The lion strain, *E. felidis* was first promoted to the species status in 2009 where it was described as a lion strain probably confined to sylvatic transmission in sub-Saharan Africa [33]. Subsequently, *E. felidis* was isolated from wildlife in Kenya [9] and South Africa [46]. This seem to confirm its adaptation to sylvatic transmission systems. Isolation of *E. felidis* in both cycles in the present study is, therefore, an indication of active interaction of wild animals and domestic dogs within the Reserve environments. It is also possible that, these domestic dogs were infected in the process of herding of livestock. During other livelihood activities such as collection of firewood, accompanying dog(s) could scavenge on wild herbivores or might have been infected through coprophagy of wild carnivorous host's fecal matter. On the other hand, wild carnivores were often observed marauding within *manyattas* (homesteads of the Samburu and Maasai pastoral communities) at night [9]. In 31 Taeniid eggs from fecal samples of wild definitive host from Samburu National Reserve (hyena, lion and side stripped jackals) were found to have *E. granulosus* s.s. Following the observations/reports of wild carnivores in *manyattas* and predation on livestock, it is highly likely that such carnivores acquired *E. granulosus* s.s. infection as a result of preying on livestock.

Transmission overlap of *E. granulosus* s.s. and *E. felidis* in the domestic and sylvatic cycles could not be fully explored with data that were available. However, our observation clearly demonstrates the interaction between domestic and wildlife definitive hosts, raising major public health concerns. The genetic proximity between *E. granulosus* s.s. and 'lion strain' *E. felidis* is well understood [34], but the pathogenic potential of *E. felidis* in livestock and human, and the importance of *E. granulosus* s.s. in wildlife intermediate hosts are some of the crucial aspects of

Cystic Echinococcosis (CE) that remain unanswered. *E. granulosus* s.s. is the most infective species to humans in locally [31], and worldwide [47]. The mere presence of *E. granulosus* s.s. in wild carnivores broadens their transmission to human and wild ungulates. So far, warthogs are known to be the most suitable intermediate hosts of *E. felidis* [34]. It is, however, unclear as to whether domestic pigs or any other livestock plays the intermediate host role to sustain the transmission of *E. felidis* in the domestic setting. Slaughter data of animals from Maasai land could not reveal the occurrence of *E. felidis* in the domestic intermediate hosts [27]. This, however, does not rule out the possibility perhaps that the disease situation in other domesticated animals in endemic wildlife vicinity might be the catalytic factor.

Echinococcus canadensis G6/7 was found in a dog fecal sample in Samburu National Reserve area. The genotype, is rare in wildlife and its existence in a domestic dog fecal sample collected from the heart of the Reserve cannot guarantee its wildlife origin. [33]. Possibly, the dog acquired the infection from the domestic setting and defecated at the heart of the Reserve during regular human visits to the area such as herding or collection of firewood mostly by Samburu Morans who often visited the heart of the Reserve often accompanied by their dogs [9]. Furthermore, in the absence of wild intermediate host, its existence in wildlife cycle can henceforth be ruled out.

Increased infection of the *Echinococcus species* and other Taenia species in domestic dogs, especially in Samburu could be due a long-standing tradition in the community where animal lungs are fed to dogs. This was supported by what the local community had to say as quoted '*specifically lungs are strictly fed to the dogs during home slaughter or/and at the abattoirs*. In the course of our study, other parasites observed included three cosmopolitan *Taenia* spp. including: *T. multiceps* and *T. hydatigena* (found in Samburu and Maasai Mara and in both cycles) and *T. saginata*, which was rare but reported here in hyena feces collected in Samburu National Reserve environment. Existence of *T. multiceps* in the domestic and wild definitive hosts is rather alarming. The parasite has earlier been reported in dogs and jackals and is known to cause severe neurological disease in animals (coenurosis) when the larva migrates to the brain and spinal cord [48]. It affects sheep and goats, being their major intermediate hosts [49]. The hyena with *T. saginata* eggs most likely acquired this through feeding on human feces as previously reported [9, 33]. This protracts the range for cattle to be infected since herding dogs interacts more closely with livestock and, therefore, may increase chances of infection for cattle.

8. Conclusion

Inadequate data on wildlife-human related infectious diseases has reduced preparedness against disease outbreak in Kenya. More studies on problems relating to wildlife diseases, determining the presence of such diseases, their prevalence and their impact on wildlife conservation and management are inevitable. Existing wildlife management systems are deficient of disease surveillance component, and this has led to human deaths and animal loses to zoonotic infectious disease. Disease transmission between the human-wildlife cycle is a generally gray area to most stakeholders, making disease management strategies difficult. Growth in human population is causing great challenges in environmental conservation management. Changes observed include wildlife habitat change, which has adversely caused ecological changes as well as increased emergence and re-emergence of zoonotic infectious diseases. Based on the findings of this study, it can be hypothesized that if proper wildlife management systems including disease surveillance systems are observed in Kenya, wild animal population will increase, the rare species will be free of illnesses, and human mortalities caused by zoonotic diseases will decrease. There is an overlap in occurrence of *E. granulosus* s.s. and *E. felidis* in wildlife and domestic settings in Kenya. Active interaction of wild and domestic *Echinococcus* in definitive hosts has been observed. However, data on importance of intermediate hosts for the ′lion strain′ *E. felidis* in domestic and *E. granulosus* s.s. in wildlife would be key in interpreting transmission dynamics of these parasites. Our study provides a base for further analysis of the sylvatic-domestic transmission interface of *Echinococcus* spp. in sub-Saharan Africa, and suggests improved wildlife conservation and management systems, with possibility of having all wildlife sanctuaries fenced, for the benefit of human and well as animals.

Acknowledgements

The authors acknowledge the support of Lynn Nkatha who assisted with field sampling and laboratory analysis. The authors further acknowledge financial assistance by cystic Echinococcosis in Sub-Saharan Africa Research Initiative.

Ethical approval

Ethical permission to conduct this research was granted by Kenya Medical Research Institute's Ethical Committee, animal care and use committee and Meru University of Science and Technology Institutional Research Ethics Review committee (MIREC-035-2017). Permission to collect samples from Maasai Mara and Samburu National Reserves was granted by Kenya Wildlife Services.

Author details

Dorothy Kagendo[1*], Eric Muchiri[1], Peter Gitonga[2] and Esther Muthoni[3]

1 Meru University of Science and Technology, Meru, Kenya

2 Kenya Wildlife Services, Nairobi, Kenya

3 Kenyatta University Hospital, Nairobi, Kenya

*Address all correspondence to: dkagendi@yahoo.com

References

[1] Anderson J F., Forcey G., Osbourne J., Spurgeon A 2002. The importance and Use of Wild life Management plans: An Example from the camp Dawson collective Training Area. West Virginia Academy of Sciences. 74

[2] Woinarski John and Fisher Alaric (2003). Conservation and the maintenance of biodiversity in the Rangelands. Rangeland Journal-Rangeland J.25.10.1071/RJ03013.

[3] Henri H., Luchman B., Batoro J (2017). The potential of Flora and Fauna as tourist attractions in Biodiversity pack of Pelewan Forest, Central Bangka. Biosaintifika: Journal of Biology and Biology Education.9.10.15294/ biosaintifika. v9i2.9225

[4] Zhou, X. H., Wan, X. T., Jin, Y. H., & Zhang, W. (2016). Concept of scientific wildlife conservation and its dissemination. Zoological research, 37(5), 270-274. https://doi. org/10.13918/j.issn.2095-8137.2016.5.270

[5] Gray Thomas., Marx Nick., Khem Vuthyrayong., Lague Dean., Nijman Vincent., Gauntlet Suwanna (2017) Holistic management of live animals confiscated from illegal wild trade; https://doi.org/10/1111/1365-2664.12916

[6] Clark Spencer Larsen, (2003) Animal source foods and human health during Evolution, The journal of Nutrition, volume 133, Issue 11, November 2003, Pages 3893S -3897S. https: doi. org/10/1093/jn/133.11.3893S

[7] Mathew, F. 2009. Zoonoses in Wildlife: integrating ecology into management. Advances in parasitology, 68, 185-209

[8] Ogutu J., Piepho H., Said M., Ojwang G., Njno L., Kifungo S., Patrick W., (2016) Extreme Wildlife Declines and concurrent increase in Livestock Numbers in Kenya: What are the causes? Published: September 27, 2016. https://doi.org/10.1371/journal. pone.0163249

[9] Kagendo, D., Magambo, J., Agola, E.L., Njega, S.M., Mulinge, E., Gitonga, P., Mbae, C., Muchiri, E., Wasserman, M., Kern, P., Roming T. 2014. A survey for *Echinococcus spp* of carnivores in wildlife conservation areas in Kenya. Parasitol. Int. 63; 604-11

[10] Moses M., and Kevin B., (2013) Factors affecting the success of conserving Biodiversity in National Parks: A Review of Case Studies from Africa. Hindawi Publishing co-corporation. International journal of Biodiversity Volume 2013, Article ID 798101, 20 pages. http://dx.doi. org/10/1155/2013/798101

[11] Kideghesho J.R. (2016). Reversing the trend of wildlife crime in Tanzania: challenges and opportunities, *Biodiversity and Conservation,* 25: 427-449

[12] Kideghesho et al. (in press). Will Tanzania wildlife sector survive the COVID-19 pandemic? Tropical Conservation Science

[13] Jingjing Yuan, Yongling Lu, Xiangui Cao & Haotian Cui (2020) Regulating wildlife conservation and food safety to prevent human exposure to novel virus, Ecosystem Health and Sustainability, 6:1, Doi:10.1080/20964129.2020.1741325

[14] Jaenson, T.G.T., Petersson, E.H., Jaenson, D.G.E. et al. The importance of wildlife in the ecology and epidemiology of the TBE virus in Sweden: incidence of human TBE correlates with abundance of deer and hares. Parasites Vectors 11, 477 (2018). https://doi.org/10.1186/ s13071-018-3057-4

[15] Vila M., Jose'E., Hejda M., Hulme P., Jaros V., Maron J., Pergl J., Schaffner U.,

Sun Y., Pys P., (2011) Ecological impacts of invasive alien plants: a meta-analysis of their effects on species, communities and ecosystems. *Ecology letters*; doi: 10.1111/j.1461-0248.2011. 01628.x; 2011, 14: 702-708

[16] World Health Organization, Regional Office for South-East Asia. (2011). Comprehensive Guideline for Prevention and Control of Dengue and Dengue Hemorrhagic Fever. Revised and expanded edition. WHO Regional Office for South-East Asia. https://apps.who.int/iris/handle/10665/204894

[17] Dennis M., Albert M., Godfrey B., Maureen K., Ann K., Grace G., Akurut., Sylvester O., Wilson A., Samuel G., Jan H., Robert T (2017) African animal trypanosomiasis as a constraint to livestock health and production in Karamoja region: a detailed qualitative and quantitative assessment; BMC veterinary Research. 3017; 13:355. DOI: 10.1186/s12917-017-1285-z

[18] Simwango M., Ngonyoka A., Nnko H., Salekwa L., Ole-Neselle M., Kimera S., Gwakisa P (2017) (Molecular prevalence of Trypanosome infections in cattle and tsetse flies in the Maasai Steppe, northern Tanzania. Parasites vectors 10. 507 (2017) https://doi.org/10.1186/s13071-017-2411-2

[19] Jingjing Yuan, Yongling Lu, Xiangui Cao & Haotian Cui (2020) Regulating wildlife conservation and food safety to prevent human exposure to novel virus, Ecosystem Health and Sustainability, 6:1, Doi:10.1080/20964129.2020.1741325

[20] John S Mackenzie and David W Smith (2020) Covid-19: a novel Zoonotic disease caused by a coronavirus from China: What we know and what we Don't: Microbiology Australia 2020. 10.1071/MA20013

[21] Tiwari R., Dharma K., Sharun K., Yatoo M., Malic Y., Singh R., Michalak I., Sah R., Aldana K., Morales A.M (2020); Covid-19: animals, Veterinary and Zoonotic links, Veterinary quarterly, 40:1, 169-182. Doi: 10.1080/01652176.2020.1766725

[22] Weiss., 2008., Zoonotic parasitic diseases: emerging issues and problems. International journal for parasitology, 38(11), 1209-210. https: doi.org/10.1016/j.ijpara.2008.05.005

[23] Kazwala R.R., (2016) Zoonotic diseases at the human- domestic interface in southern and Eastern Africa International Journal of Infectious Diseases 53S (2016) 4-163; Doi: https://doi.org/10.1016/j/ijid.2016.11.016

[24] Romig, T., Omer, R.A., Zeyhle, E., Hüttner, M., Dinkel, A., Siefert, L., Elmahdi, I.E., Magambo, J., Ocaido, M., Menezes, C.N., Ahmed, M.E., Mbae, C., Grobusch, M.P., Kern, P., 2011. Echinococcosis in sub-Saharan Africa: emerging complexity. Vet. Parasitol. 181, 43-47

[25] Mbaya, H., Magambo, J., S. Njenga., Zeyhle. E., Mbae. C., Mulinge. E., Wassermann. M., Kern, P. and Romig, T. 2014. "Echinococcus spp. in central Kenya: a different story." Parasitol. Res. 113(10): 3789-3794

[26] Wassermann, M., Aschenborn, O., Aschenborn, J., Mackenstedt U., Romig, T., 2015. A sylvatic lifecycle of Echinococcus equinus in Etosha National Park, Namibia. Int. J. Parasitol: Parasites of wildlife 4. 97-103

[27] Addy, F., Wassermann, M., Kagendo, D., Ebi, D., Zeyhle, E., Elhahdi, I., Umhand, G., Casuli, A., Harandi, M., Aschenborn, O., Kern, P., Mackenstedt, U., Romig, T., 2017. Genetic differentiation of the G6/7 cluster of Echinococcus canadensis based on mitochondrial marker genes. Int. J. Parasitol., 47.923-931

[28] Cardona G., Carmena D; 2013 A review of the global prevalence,

molecular epidemiology and economics of cystic echinococcosis in production animals, Veterinary Parasitology, Volume 192, Issues 1-3 Pages 10-32 Issn 0304-4017 https://doi.org/10.1016/j.vetpar.2012.09.027.

[29] Alvarez, R., Romig, T., Lighttowlwers, M.W., 2014. Echinococcus granulosus sensu lato genotypes infecting humans-review of current knowledge. Int.j. parasitol,44,9-18

[30] Okello, GBA., (1986). Hydatid disease: Research and control in Turkana III. Albendazole in the treatment of inoperable hydatid disease in Kenya report on 12 cases. Transactions of the royal society of Tropical Medicine and Hygiene. 1986:193-1957

[31] Mutwiri, T., Magambo, J., Zehyle, E., Mkoji, G.M., Wamae, C.N., Mulinge, E., Mbae, C., Wassermann, H., Kern, P., Romig, T. 2013. Molecular characterization of Echinococcus granulosus species/strains in human infections from Turkana, Kenya. East. Afr. Med. J. 90, 235-240

[32] Magambo, J., Njoroge, E., Zehyle, E., (2006). Epidemiology and control of Echinococcosis in sub-Saharan Africa. Parasitology International. 55: S193-95

[33] Mulinge, E., Magambo, J., Odongo, D., Njenga, S., Zeyhle, E., Mbae, C., Kagendo, D., Addy, F., Ebi, D., Wasserman, M., Kern, P., Romig, T., 2018 Molecular characterization of Echinococcus species in dogs from four regions of Kenya. Vet. Parasitol. 255; 49-57

[34] Hüttner, M., Siefert, L., Mackenstedt, U., Roming, T. 2009. A survey of Echinococcus species in wild carnivores and livestock in East Africa. Int. J. Parasitol. 39; 1269-76

[35] Stuart & Stuart, (2000). A Field Guide to the Tracks & Signs of Southern and Eastern African Wildlife, Cape Town, 3rd Edition. Struik Publishers

[36] Mathis, A., Deplazes, P., Eckert, J., 1996. Improved test system for PCR based specific detection of Echinococcus multilocular is eggs. J. Helminthol. 70, 219-222.

[37] Lucy Wamuyu., Vincent Obanda., Daniel Kariuki., Francis Gakuya., Moni Makanda., Moses Otiende., and Sheila Ommer (2015) Molecular detection and characterization of Theileria Infecting Wildebeest (*Connochaetes taurinus*) in the Maasai Mara National Reserve, Kenya; Department of Biochemistry, Jomo Kenyatta University of Agriculture and Technology (JKUAT), Pathogens 2015, 4(3), 626-638; https://doi.org/10.3390/pathogens4030626

[38] Haddad, N. M., Brudvig, L.A., Clobert, J., Davies, K.F., Gonzalez, A., Holt, R.D., Lovejoy, T.E., Sexton, J.O., Austin, M.P., Collins, C.D., Cook, W.M., Damschem, E.I., Ewers, R.M., Foster, B.L., Jenkins, C.N., King, A. J., Laurance, W.F/. Levey, D.J., Margules, C.R., Melbourne, B.A., ... Townshend, J.R., (2015). Habitat fragmentation and its lasting impact on Earth's ecosystems. Science advances, 1(2), e1500052. https://doi.org/10.1126/sciadv.1500052

[39] Reusken C.B., Leven M., Sigfrid L., Eckerle L., Koopmans M. / (2018) Laboratory preparedness and response with a focus on arboviruses in Europe; Clinical Microbiology and Infection 24 (2018) 221e228

[40] Mohamed E. El Zowalaty, Sean G. Young & Josef D. Järhult (2020) Environmental impact of the COVID-19 pandemic_ a lesion for the future, infection ecology & Epidemiology, 10:1, DOI: 10.1080/20008686.2020.1768023

[41] Mcelwain, T.F., &Thumbi, S.M (2017). Animal pathogens and their impact on animal health, the economy, food security, food safety and public

health. Revue Sceintifique et technique (International Office of Epizootics), 36(2), 423-433. HTTPS://doi.org/10.20506/rst.36.2.2663

[42] Wambua, L., Wambua, O.N., Ramogo, A.M., Mijele, D., &Otiende, M.Y (2016). Wildebeest-associated malignant catarrhal fever: perspectives for integrated control of a lymphoproliferative disease of cattle in sub-Saharan Africa. Archives of virology, 161(1), 1-10. https://doi.org/10.1007/s00705-015-2617-6

[43] Conover, Michael & Pitt, William & Kessler, K. & DuBow, T. & Sanborn, W. (1995). Review of human injuries, illnesses, and economic losses caused by wildlife in the United States. Wildlife Society Bulletin. 23. 407-414.

[44] Odero, J., Magambo, J., Zehyle, E., Kutima, H., Ndahi, L., Njoroge, F., Mutwiri, T., Mbaya, H., Mbae, C., Mulinge, E., Kagendo, D., Kern, P., Romig, T., (2014). The monetary burden of cystic echinococcosis and its distribution based on gender and age in Turkana North District, Kenya. Asian Journal of Pharmaceutical and Health Sciences: 4: 1135-1142

[45] Stoldt, M., Göttert, T., Mann, C. et al. Transfrontier Conservation Areas and Human-Wildlife Conflict: The Case of the Namibian Component of the Kavango-Zambezi (KAZA) TFCA. Sce Rep 10,7964(2020). https://doi.org/10.1038/s41598-020-64537-9

[46] Halajian, A., Luus, Powell W.J., Roux, F., Nakao, M., Sasaki, M., Lavikaieni, A., 2017. Echinococcus felidis in hippopotamus, South Africa. Vet. Parasitol.243, 24-28.

[47] Alvarez, R., Romig, T., Lighttowlwers, M.W., 2014. Echinococcus granulosus sensu lato genotypes infecting humans-review of current knowledge. Int.j. parasitol,44,9-18

[48] Amrabadi, O., Oryan, A., Moazen, M., Sharifiyazdim, H., Akbari, M., 2015. Comparison of cerebral and non-cerebral coenurus by genetic markers of glycolytic enzyme (enolase) and mitochondrial sequences in sheep and goats. Vet. Parasitol. 15:214 (3-4):333-6 Doi10.1016/J.Vetpara.2015.10.021

[49] Amer S., Elkhatam A., Fukuda Y., Bakr L., Zidan AE., Mohamed MA., Tada C., Nakai Y. 2017. Clinical, pathological and molecular data concerning coenurus cerebralis in sheep in Egypt. Published online doi: 10.1016/j.dib.2017.10.070 16(2018)1-9

Chapter 6

Wildlife Management Areas in Tanzania: Vulnerability and Survival Amidst COVID-19

Rehema Abeli Shoo, Elizabeth Kamili Mtui, Julius Modest Kimaro, Neema Robert Kinabo, Gladys Joseph Lendii and Jafari R. Kideghesho

Abstract

The establishment of Wildlife Management Areas (WMAs) has been adopted as intervention to safeguard the wildlife and their habitats outside the core protected areas in Tanzania. Along with their conservation role, WMAs provide an opportunity for local communities to derive economic benefits from wildlife-based enterprises on their land. WMAs primarily rely on revenues generated from photographic and hunting tourism to support operational activities and create incentives for the local communities to conserve wildlife resources. The current global travel restrictions and lockdown caused by an outbreak of COVID-19 pandemic have reduced a vital funding source for WMAs. This, therefore, undermines the ability to manage the wildlife resources and reward communities for the opportunity cost of their land and other costs associated with coexisting with wildlife. This chapter examines the extent to which the decline of tourism revenues as a result of the outbreak of COVID-19 pandemic has affected WMAs as a framework for local communities to manage and benefit from wildlife. Data were collected through semi-structured interviews on five WMAs in Northern Tanzania that were purposively selected based on their ability to generate a significant amount of revenues from tourism. Findings show that the decline of tourism revenues triggers unprecedented adverse effects on the conservation of wildlife resources within WMAs. Livelihood of the local communities is also affected due to loss of employment opportunities and drop-off of tourism income obtained from the sales of local goods to the tourists and tourist hotels. We recommend the creation of local mechanisms for revenue acquisition that are more resilient to global shocks, diversifying revenue-generating options within WMAs, and putting in place the right funding model that would warrant WMAs sustainability.

Keywords: Wildlife Management Areas, COVID-19, local communities, tourism, livelihood, conservation

1. Introduction

Recognition of the close link between sustainable natural resource management and rural development and a pervasive decline of wildlife due to traditional centralized wildlife management practices prompted the introduction of Community-Based Wildlife Management (CBWM) in Tanzania [1–3]. CBWM approach intends to fill the gaps emanating from 'protectionist' wildlife management practices whereby local communities were strictly excluded from utilization and management of wildlife [4]. The main assumption of CBWM approach is that access of economic benefits from wildlife-related enterprises among the local communities improves their standard of living and, consequently, motivate them to support wildlife conservation efforts [5–7]. The endorsement of Wildlife Policy in 1998 laid out the initial underpinnings of Tanzania's approach to CBWM through the establishment of WMAs. The concept "represents an important option for wildlife conservation outside the core protected areas and acts as a buffer against human impacts while enhancing rural economic development" [8]. These community-managed areas are established within the village lands, whereby groups of villages set aside land for sustainable wildlife conservation and derive benefits from investment options provided by the land [9].

Despite the governance and conservation challenges undermining the success and sustainability of WMAS [5, 10, 11], WMAs are thought to represent a feasible option for conserving wildlife outside the core protected areas in Tanzania while enhancing rural economic development [12, 13]. A 2012 status report on WMAs by WWF, highlights some evidence that the creation of WMAs has led to improved biodiversity conservation and increased protection of areas that are considered ecologically important [5]. WMAs have also given villages a framework within which they can better manage their land and benefit from the wildlife through tourism activities [11].

Literature shows that tourism in community lands outside the core protected areas, particularly in northern Tanzania, has expanded and many villages earn considerable revenues through joint ventures [14–16]. By forgoing land uses such as farming, livestock grazing or settlement, and allowing investors to use the land for photographic and hunting tourism, the villagers receive a set of payments. Currently, several options for utilizing and creating benefits from wildlife resources exist in the WMAs. These include non-consumptive (photographic tourism) and consumptive (trophy hunting, resident hunting) uses that are carried out in a manner that does not interrupt the ecological processes and the ecosystem functions. For example, tourist lodges in IKONA and Burunge WMAs generate over one million US$ per annum (See **Table 1**). Concessions are shared among the villages and used to fund the development and maintenance of social infrastructures like schools, dispensaries, or water projects. These benefits have, consequently, instilled positive attitudes towards conservation among the villagers, who are actively taking part in conservation through patrols, intelligence, and de-snaring [12]. Besides generating revenues, tourism development in WMAs provides local employment and market for local goods such as foods, souvenirs, and handicrafts. The income

WMA	District	Revenues from Hunting	Revenues from Photographic Tourism	Total revenues
1. IKONA	Serengeti	735,510	933,076	1,668,586
2. Burunge	Babati	78,261	1,189,785	1,268,046
3. Makao	Meatu	14,861	N/A	14,861
4. Enduimet	Longido	117,834	132,799	250,633
5. Randilen	Monduli	N/A[1]	332,801	332,801

Source: Respective Authorized Associations.
[1]*N/A: No operation of the respective activity.*

Table 1.
Revenues (US$) from hunting and photographic tourism disbursed to five WMAs of northern Tanzania from 2016/2017 to 2017/2018.

from tourism represents a growing source of economic diversification, poverty alleviation as well as economic incentives for community stewardship towards wildlife resources. Community Conservation Banks (COCOBA), also known as village savings and loan groups, have emerged as a popular strategy in guaranteeing the local livelihoods and the future of wildlife populations found on village lands outside the core protected areas [17]. Through the strategy, the communities are supported financially by allowing them to invest in environmentally-friendly businesses and, thus, lessening the dependence on ecologically damaging activities such as bushmeat hunting, farming, or cattle grazing.

1.1 COVID-19 pandemic, tourism and WMA

In December 2019, the Corona Virus Disease (COVID-19), caused by a Severe Acute Respiratory Syndrome CoronaVirus2 (SARS-CoV-2), was first reported in Hubei Province, which is a famous animal market in the Chinese city of Wuhan [18]. On 30 January and 11 March 2020, the disease, was declared 'A Public Health Emergency of International Concern' and 'a pandemic', respectively [19]. The anthropogenic impact on nature and biodiversity loss through deforestation and the modification of natural habitats have been associated with the outbreak of this pandemic [20–22].

The COVID-19 pandemic has pushed the world to a devastating economic recession, the tourism industry, being one of the hardest hit [15, 16, 23]. Lockdown and travel restrictions imposed around the world due to this pandemic have halted operations in the tourism industry as a result of the decline of cash flow from incoming tourists [24]. The World Tourism Organization (UNWTO) assessment of the economic impact of the pandemic on international tourism estimated a loss of US$300–450 billion in 2020 compared to 2019 figures [25]. Monthly surveys conducted by SafariBookings.com from March to October 2020, involving hundreds of safari tour operators reported an over a three-quarter decline of bookings and massive cancelations of the bookings [26]. Also noted in these surveys, was a failure among the operators to meet their administrative costs [17].

Loss of revenues from tourism poses a severe threat to the survival of protected areas as these revenues are reinvested for the management of wildlife species and habitats. Tourism revenues are crucial for funding conservation operations and support the benefit sharing schemes among the communities. Furthermore, these revenues enable conservation as a land-use to compete effectively with other economic activities that are ecologically damaging. WMAs, like many other categories of wildlife protected areas in Tanzania, rely entirely on tourism revenues and partly on conservation donors for their operations. The collapse of tourism, therefore, increases their vulnerability to the global impacts of COVID-19. The COVID-19 pandemic is likely to strip off the vital funding for WMAs needed to manage the wildlife resources and reward communities for the opportunity cost of coexisting with wildlife. Tourism income contributes to almost 90–95% of conservation management costs in WMAs. The WMA's administrative costs include the remuneration of the staff - mainly community rangers, purchase of equipment and supplies needed for rangers to remain active in combating illegal uses of wildlife resources. Without adequate funding, the operational budget for anti-poaching surveillance and other activities for WMAs will seriously be affected. Further to impact on WMA operations, compromised livelihoods options as a result of reduced tourism incomes will likely intensify threats to wildlife resources through increased illegal activities pursued to cope with economic hardship. Evidence-based literature indicates that poverty is one of the critical drivers of wildlife and environmental crimes in Africa [27–31].

Considering the importance of WMAs in securing wildlife space around the core protected areas and creating incentives for local communities to conserve wildlife resources, it is apparent that the failure of these areas to function properly will bear the negative repercussions on the local livelihoods and biodiversity. Understanding how WMAs can sustain their vital roles of conservation and improving the livelihoods of the vulnerable communities during the current unprecedented circumstance of COVID-19 is critical for their survival and management. This paper, therefore, analyses the extent to which the collapse of tourism activities due to COVID-19 pandemic has affected WMAs operational activities and local livelihoods. Our findings will contribute to further understanding of the effects that contemporary issues such as infectious diseases and, COVID 19, in particular, have on conservation, tourism, and local livelihoods. These findings are quite robust in recommending possible mechanisms of enhancing resilience and preparedness of community-managed conservation areas such as WMAs to emerging global pandemics.

2. Methodology

2.1 Study area

The study was conducted in the five purposively selected WMAs, namely; Makao, Burunge, Enduimet, Randileni, and IKONA located in Northern Tanzania (**Figure 1**). These five WMAs along with 33 others, covers a total area of 29,000 km^2 which is about 3% of the country's land surface area. Details of these WMAs are shown in **Table 2** below.

The selection of the five WMAs was based on their performance in generating revenues from tourism activities compared to other WMAs in the country. For example, from 2016 to 2018 hunting and photographic tourism earned these WMAs a total of US$ 3.55 million (US$0.95 million from tourist hunting and 2.6 million from photographic tourism [32].

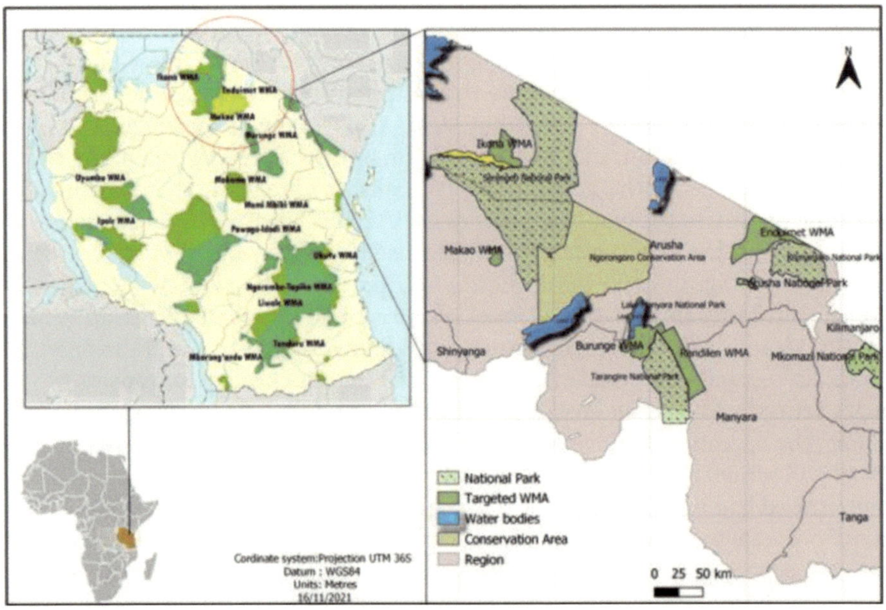

Figure 1.
Location of selected WMA's in the northern Tanzania.

WMA	District	Size (Km^2)	Year of establishment	Participating villages	GN and Date of gazettement
IKONA	Serengeti	242.3	2003	5	GN 57 (09.03.2007)
Burunge	Babati	617.0	2003	10	GN 37 (31.06.2006)
Makao	Meatu	768.9	2007	7	GN 369 (20.11.2009)
Enduimet	Longido	751.4	2003	9	GN 57 (09.03.2007)
Randilen	Monduli	312.0	2011	6	GN 21 (01.02.2013)

Table 2.
Status of selected wildlife management areas.

2.2 Study participants and sampling method

The study population consisted of Village Game Scouts (VGS), village government leaders, leaders of the WMAs and ten community members residing in each of the five selected WMAs. The WMA leaders were purposively selected based on their positions and roles they play in the management of WMAs. The community members consist of villagers who have no leadership role in the village governments or WMAs but who derive benefits from tourism activities taking place in the WMAs, directly or indirectly. These participants (**Table 3**) were selected using both purposive and snowball sampling methods, considering their diverse roles in the community as far as WMAs, conservation, and tourism are concerned. In addition, with purposive sampling technique, efforts were made to consider gender of respondents considering different roles that males and females play in conservation and tourism.

2.3 Sources of data and methods for data collection

This study was conducted in May, 2020 and employed both primary and secondary data to generate information on the effects of COVID-19 on survival of WMAs and impacts on local livelihoods. Secondary data from different reports complemented the primary data obtained through interviews with key respondents. The use of qualitative data collection methods, and in this case, key informant interviews was deemed appropriate in providing a detailed understanding of WMAs performance amid the COVID-19 pandemic. Key informants were anticipated to provide a broader understanding of socio-ecological dimensions of the current situation facing WMAs in Tanzania and capture the complexities as well as diverse perspectives as the situation unfolds. In particular, key informant interviews with selected community members sought information on how tourism influences their livelihood and how the collapse of tourism activities in the WMAs has been affecting their socio-economic development.

Categories of participants	Wildlife Management Areas				
	IKONA	Burunge	Makao	Enduimet	Randilen
1. Village Game Scouts (VGS)	3	2	4	1	1
2. Village Leaders	2	2	2	2	2
3. WMAs Leaders	1	1	1	1	1
4. Community members	10	10	10	10	10

Table 3.
Number of participants in each category of respondents in the WMAs.

On the other hand, WMAs leaders provided information on tourism revenue sharing mechanisms, the local support for conservation efforts as a result of tourism, and how the decline of tourism activities within each selected WMA has affected both the WMAs' operational activities and local livelihoods. A checklist of questions was designed to guide an interview based on the key research objectives. The interviews were conducted conversationally to allow probing of the issues which were not included in the interview guide but useful in addressing the research questions. Social distance and other precautionary measures to avoid infection of COVID-19 were observed during the interview.

2.4 Data analysis

Content analysis was employed to analyze data from the interviews after translation from Swahili to the English language. Content analysis is a procedure for analyzing textual material (opinions and perspectives) which involves compressing texts into themes and content categories to uncover frequency of occurrence among texts [33, 34]. In this study, the content of the interview data was classified into important themes that were identified to reflect the objectives of the study. Each theme presented an "idea" and all the data related to a particular theme were added under that unit. The themes were evaluated by making comparisons of responses and determining how often each of the categories appears to conclude a specific aspect. The presentation of the results is also supported by quotes from respondents' responses for illustration purposes.

3. Results and discussion

3.1 Income generation from wildlife management areas

Results from the interviews show that tourism is an important income-generating activity in the five WMAs. Essentially, about 90% of WMAs income is obtained through various tourism-related opportunities such as photographic safaris, hunting tourism, concession fees, and other fees. The remaining 10% come from donor support in the form of grants and occasionally from penalties in form of fines imposed against various law offenders within the WMAs. The financial contribution of hunting and photographic tourism to five WMAs is corroborated by the 2018 Wildlife Sub-sector Statistical Bulletin [32].

The revenue from the WMAs is shared among the Central Government, Local Government and the respective WMAs. The participating villages in the WMAs receive 50% WMA's income share from the Government and the rest is used to cover the operational costs. Besides direct revenues, tourism has created numerous opportunities to diversify the local economy. It is important to note that livelihood diversification in WMAs is essential, as most of these areas are on lands that are considered economically marginal, experiencing semi-arid to arid conditions with limited possibilities for generating household income from other activities such as crop farming and livestock production. Tourism in such areas has been widely acknowledged to present opportunities and capitalize on the cultural, landscape, and wildlife assets commonly found on marginal lands that are not particularly valuable for other activities that drive the economy [35].

Further analysis of the income opportunities in the WMAs shows that tourism provides, not only a regular source of income to the member villages, but also adds value to different locally produced products in the course of tourists spending (**Table 4**). There were several ways through which the local communities in the

Categories of opportunities	Description
1. A partnership between investors and communities	• Annual concession fees from contracts made between the WMA member villages and investors • Tourist lodges and hotels in the WMA employing local staff
2. Locally controlled small enterprises	• Community members selling cultural/ traditional products and handicrafts to tourists
3. Local services to tourists and lodges	• Local tour guiding • Supplying agricultural-based products, vegetables, and meat to tourist lodges • Performing traditional dances to entertain the tourists
4. Income share from WMA to local communities	• The participating villages receive 50% of WMA's income share after Government deductions

Table 4.
Opportunities for communities to benefit from WMAs.

WMAs village members have been able to benefit from the presence of tourists in their areas. For example, it was revealed that some community members could derive benefits through working as local tour guides, direct employments in the tourist lodges and hotels as well as selling a range of locally made products such mats and baskets, beaded necklaces, earrings and bracelet, and Maasai attires to the tourists. It was revealed that, on average, community members engaging in selling local products and/or providing services to tourists might earn up to TZS 300,000 (approximately USD 130) per month. The local communities have also been hired occasionally by the tourist lodges and hotels to entertain tourists through performing traditional dances, where on average; each earns about 50,000 Tanzanian shillings per performance. There is a general understanding that tourism income within the WMAs positively influences the livelihoods of those who directly engage in tourism-related activities. Hence the collapse of tourism has affected the livelihood of several members of the communities who have not been able to derive any income from such activities for the past 6 months. One of the respondents working as a Village Game Scout (VGS) in Burunge WMA underscored the importance of tourism income by reiterating that;

"... It is through this job that I was able to build a decent brick house for my parents and I am capable of paying for my children's school fees ... ".

Tourism-related income in the WMAs was also found to directly influence households' access to basic education, health care, and other social services for the community members who do not have direct engagement with tourism activities. Results from interview indicate that the majority of the members of the communities benefit through collective income earned at the community level. This collective income is usually obtained either through WMAs' share to the communities or when villages earn profits through leasing part of their land to tourism business investors. The communal income is usually invested in improving the provision of social infrastructure such as water, schools, roads, and dispensaries within the community. Some tour operators have been offering both financial and material support to support the communities rehabilitating school buildings as well as purchasing equipment required for schools and health care operations. Community development projects have generally been the fundamental means of extending tourism benefits to all community members. The funding of social services projects through tourism revenues relieves the communities from the burden of contributing to these services, thus making the money available to cater for alternative household requirements.

3.2 Effects of COVID-19 pandemic on WMAs' operational activities and wildlife resources

Discussion with leaders of the WMAs uncovered that, following an outbreak of COVID-19 pandemic and, subsequently, loss of revenues from tourism, none of the five WMAs was able to effectively implement its administrative activities including the funding of the village development projects. For example, the Enduimet WMA was forced to defer implementation of the infrastructure development projects that were planned during the 2019/2020 financial year, among others, being the renovation of the ranger posts, construction of game viewing routes and a visitor centre in Engikaret. Other administrative activities that were put on hold include the review of village land use plans, tree planting activities, environmental conservation education through cinemas, and the Annual General Meeting (**Table 5**).

BURUNGE • Failure to put in place a Spatial Monitoring and Reporting Tool (SMART) System for data collection and anti-poaching activities • Failure to purchase field gears for the scouts and crop protection tools • 24 local staff working in the hotels within the WMA were laid-off • Various village development projects in the WMA member villages were put on hold • Small scale businesses such as the sale of souvenirs closed • 50 local tour guides in the WMA lost their jobs • A planned improvement of the road networks within the WMA stopped • Could not conduct environmental education program to the local communities • Overfishing by the local communities within the WMA
ENDUIMET • 15 employees laid off • Anti-poaching operations are confined within the WMA Borders, very little or no extension beyond borders and neighbors • Failure to implement social development projects such as water, education, and health infrastructures within the participating villages • Failure to develop and renovate the planned tourism facilities such as tented camps, Visitor Center in Engikaret, Game viewing Routes, and the Ranger posts. • Failure to undertake and review the village land-use plans • Failure to implement conservation projects and plans including tree planting, conservation and environmental education through Cinema
RANDILEN • The collapse of local businesses especially those selling souvenirs for the past 3 months • Traditional dancers have no jobs • The number of patrols has decreased due to a lack of funding Village development projects have been put on hold
IKONA • Nearly all local tour guides have lost their jobs • 26 employees who have been paid by the village given leave without pay • Village Game Scouts (VGS) have been reduced in number from 25 to 14 • VGS undertake patrols without payment, they are just given food and no payments • Many activities like cultural activities are currently implemented • Closure of tourists hotels and local Business closed • Failure to fund development projects and support students who have been sponsored the villages
MAKAO • VGS are temporally given leave without pay while others are working on loan • Lack of donation from the tour operators • Road construction project has stopped • Reduced numbers of patrols, giving rise to increasing of illegal activities

Table 5.
Negative effects of COVID-19 on WMA resources and local livelihoods.

In Burunge WMA, the planned projects that were affected included the development of the well-equipped permit infrastructures at the WMA common entry gate; establishment of a smart system database for patrol, wildlife monitoring, and data collection as well as the purchase of crop protection tools and field gears for the scouts. Other plans which are deferred pending returning of the situation to normality are the construction of different tourism user facilities like picnic sites and viewpoints, improvement of the road networks to stimulate tourism activities within the WMA, and implementing Conservation Education programs to local communities (**Table 5**).

The effects of COVID-19 have also been notable in the anti-poaching operations. For instance, IKONA WMA has been forced to retrench 11 out of its 25 village game scouts alongside reducing the number of patrol days. Furthermore, the remaining scouts have been working without pay and are either facing reduced salaries or unpaid leave. In Randilen WMA, no payment is offered to scouts participating in patrols other than meals. In Makao WMA, 15 scouts participate in patrols work on a rotational basis and receive no pay. The anti-poaching operations in Enduimet WMA are currently being confined within the WMA borders, with no extension beyond the borders (**Table 5**). Inadequate workforce to patrol the entire WMA areas and surrounding areas along with reduced working morale among the rangers due to minimal incentives, has translated into inefficient anti-poaching operations and, therefore, render the WMAs and surrounding areas vulnerable to poaching and other illegal activities. On the other hand, loss of income as a result of retrenchment and interruption of livelihood strategies prompts illegal use of resources. The effects of COVID-19 on conservation operations in WMAs are illustrated by views from some WMA leaders provided in **Box 1** below.

"…A steep decline of tourism due to COVID-19 has a devastating effect on the cash flow to WMA. The purchasing power of the WMA has drastically declined to the extent that no procurement is currently being made."

Chairperson, IKONA WMA

"…The layoff of one-third of employees who were mainly responsible for carrying out anti-poaching operations has intensified pressure on natural resources. A WMA cannot do frequent patrols as there are no funds for fuel, vehicle repairs, and allowances for rangers. A few number staff are required to patrol a relatively large chunk of land and, thus making the anti-poaching operations a huge burden to WMA … "

Chairperson, Enduimet WMA

"…Loss of jobs, reduced salaries, and unpaid leaves have caused food insecurity among the members of the communities who are currently resorting to poaching and destruction of wildlife habitat if they are not assisted immediately. If this situation persists, I am afraid it might lead to overexploitation of the resources that attract tourists … "

Chairperson, Burunge WMA

Box 1.
Opinions of WMA leaders on effects of COVID-19 on wildlife conservation.

Patrols in WMAs are essential in monitoring wildlife, preventing poaching, and in minimizing human-wildlife conflicts. Effective implementation of anti-poaching patrols in terms of the adequate number of personnel is, therefore, essential in providing an effective deterrent against illegal activities in an area [36].

3.3 Effects of COVID-19 crisis on local livelihoods

Assessment of the effects of the COVID-19 crisis on local livelihoods was crucial in understanding the livelihood losses due to the decline of tourism activities in the

WMAs. This is important due to direct and indirect correlations between the loss of livelihood options and destruction of wildlife habitats and species [28, 30, 31].

The findings of this study show that COVID-19 pandemic has had far reaching consequences on the livelihoods of the local communities. All WMAs participating villages have failed to implement various social development projects such as water, education, and health infrastructure, to mention a few, as there are no more dividends to finance these projects. The pandemic has also led to the temporary closure of all businesses that entirely rely on tourists for revenues. It was noted that each surveyed WMA had at least one hotel investment all of which have been closed down following the pandemic, a situation that has made many locals working as casual employees for the hotels to lose their jobs.

Loss of livelihood is further contributed by the closure of small scale local businesses that were conducted by the majority of community members. It was found that the owners of these small scale enterprises were unable to run their businesses due to a lack of tourists and the consequential diminished demand for their products. Other respondents reported that they had debts and loans which they could hardly service due to the impact of COVID-19. **Box 2** below summarizes some respondents' views on impacts that have brought by COVID-19 on their livelihoods.

"...Villagers who were employed by tourist hotels are now at home without any jobs. As a consequence, the tourism multiplier effect in society has dropped drastically due to the poor performance of other economic activities ... "

Nata Village Executive Officer, IKONA WMA

"...I used to supply vegetables and goats and sheep meat to the Singita-Grumeti Reserves Hotel Ltd. and earn at least one million Tanzanian Shillings per week from selling such products, but currently, I earn nothing because the business has collapsed ... " .

A villager of Robanda, IKONA WMA

"..I used all the money I had to buy materials for making handicrafts anticipating profits from selling such products to the tourists. Currently, I am just at home left with no money or business. If this situation continues for a long period, I will not be able to cater for my household basic needs..."

Member of the Cultural Women Group in Enduimet WMA

"...I took a loan from the Village Community Bank to buy raw materials for making handcrafts ... but I am unable to repay the loan because over the past three months I have not sold a single item that I made for tourists..."

A respondent, Enduimet WMA

Box 2.
Respondents opinions on effects of COVID-19 on local livelihoods

Tanzania Chamber of Commerce, Industry and Agriculture (TCCIA) indicate that Small and Medium Enterprises (SMEs) in the country comprise 95% of the businesses in the country and contribute about 35% of the country's GDP. Compared to more prominent firms, SMEs are unlikely to survive economic shocks such as COVID-19 due to limited resources. The high concentration of SMEs in the tourism industry including their multiplier effects translate into disruption of value chains and, consequently, economic hardship to the majority of the people. A survey conducted in April 2020 by Launchpad Tanzania and Maarifa Hub to determine the severity of the economic impact of COVID-19 on small and medium businesses as well as entrepreneurs in Tanzania revealed that 97% of those surveyed had been economically impacted by the pandemic whereby 40% have experienced a loss of half of their income and 10% have experienced a total loss of income [37]. The tour companies dealing with hotel operations in the surveyed WMAs have also reported crises caused by COVID-19 due to the ongoing accommodation cancelations without payment of cancelation fees coupled with a drop in future booking requests. It is important to note that the lodge and hotel establishments in the WMAs have created several job opportunities, the majority of which are held by members of the communities. This was attested by the campsite manager for IKONA Bush Camp in IKONA WMA with a viewpoint that;

"... each staff working in our lodge has at least ten dependents. This means hundreds of individuals and families across the community are supported by jobs created by our lodge. The layoffs of staff and unpaid leaves have devastating effects on the livelihoods of the majority of members of the community ... ".

4. Implications of the study findings

It is apparent that COVID-19 has set off unprecedented crisis in the tourism economy given the immense shock to the sector [38]. Findings of this study indicate that the pandemic has posed significant negative repercussions to the majority of local communities through interruptions of their livelihood options. This presents a potential threat to wildlife resources in the surveyed WMAs. As pointed earlier in the study findings, conservation of wildlife in WMAs largely depends on tourism revenue, and hence loss of such revenue due to pandemic has far-reaching implications for local livelihoods and wildlife conservation in particular. It is widely documented that communities living adjacent to core protected areas bear the highest cost of conservation through a loss of their land and human-wildlife conflicts [39, 40]. Loss of tourism revenues has not only increased opportunity costs of conservation to the local communities but also poses a significant threat to WMAs functioning as an alternative conservation approach and a reliable land-use option.

Furthermore, it is estimated that around 80% of all tourism businesses in Tanzania are informal SMEs characterized by low-skilled individuals and vulnerable segments of the population who are either employed or run micro and small enterprises [41]. The contribution of SMEs in creating employment is widely acknowledged globally, and in Tanzania in particular [42–44]. SMEs constitute over 90% of the businesses in Tanzania and are regarded as the engine for economic growth in the country [45]. Unlike large enterprises, SMEs can be easily established since their requirements in terms of capital; technology, management and even utilities are less demanding. However, when it comes to shocks, such as pandemics, the informal workforce engaging in SMEs bears the highest vulnerability, due to lack of safety nets in terms of limited social protection measures and inadequate savings.

Local communities within the surveyed WMAs experience a similar situation because their household income is barely diversified beyond tourism, and the majority of community members are either employed or self-employed in the informal tourism economy. The decline of tourism in the WMAs has left many households vulnerable due to loss of jobs (for employees) and closure of small businesses as a result of reduced demand for their products. It should further be noted that the closure of hotel businesses within the WMAs, has not only affected the workforce who were laid off and or given leave without pay, but also had a far-reaching impact on families who were supplying vegetables and meat and poultry products to the hotels. The crisis in the WMAs is also believed to have reduced household savings with only very few households having enough savings to meet their household expenses for one month or more. Inadequate savings among the community members intensify the extent of vulnerability for many households in the study area.

Since tourism is the main income-generating activity in the WMAs, its suspension following the outbreak of COVID-19, will bear the major ramifications for wildlife conservation in the country. Our findings suggest that the pandemic has posed a major threat to conservation in the WMAs and hampered the conservation-related activities in numerous ways. Most of the conservation and development programs that were already planned by the WMAs have become uncertain (**Table 5**). The fact that the five WMAs studied are prototypes representing a success story for other WMAs existing or proposed in the country, there is a risk that the whole concept of WMA may become unpopular if the pandemic results to a total failure in these areas, which have already won the hearts of the communities. This will water down the conviction that wildlife conservation can compete effectively with alternative land uses and, thus, prompt people to claim back their land for these alternative uses, which essentially are ecologically destructive.

Community conservation programs are essential in motivating the local communities to conserve resources and pursue activities that are consistent with the preservation of nature. The drop in tourism revenues and the resultant layoffs of key staff implies inadequate resources for WMAs to conduct effective anti-poaching activities, monitor wildlife, and respond to human-wildlife conflicts. The COVID 19 crisis is also likely to intensify poverty and food insecurity among community members due to the decline of tourism income. In such cases, food-insecure communities are likely to resort to illegal natural resource for the sake of surviving. Additionally, since some WMAs partly depend on donor support for their operations, the dwindling economies may reduce the capacity of donors to provide financial support for wildlife conservation. Reduced donor support will, to a certain degree impair conservation operations by limiting the ability of WMAs to conserve and manage wildlife resources.

5. Survival mechanisms during and after the COVID-19

Addressing the effects of COVID 19 is critical in ensuring the sustainability of WMAs. Here, we describe actions necessary to sustain wildlife conservation and the livelihood of the vulnerable communities in an attempt to stir up further debate on the sustainability of WMAs in Tanzania as a framework for communities to manage and benefit from wildlife resources.

The volatility nature and the over-reliance on international tourism to support conservation and local livelihood subject WMAs to an increased degree of vulnerability to effects of global pandemics such as COVID-19. Creating local revenue mechanisms that are more resilient to global shocks may guarantee the long-term survival of WMAs against the negative effects of the pandemics. Such mechanisms, among others, include the need to expand and promote domestic tourism to locally boost wildlife-based tourism revenues and foster long-term community support for conservation. Currently, Tanzania receives about 1.5 m tourists annually, and domestic tourism represents only 26% of the total number [46]. Domestic tourism when properly developed and managed has the potential to create a more resilient tourism income and, can be a more reliable form of income for vulnerable communities as it does not fluctuate as much compared to international tourism. Among the factors that need to be addressed in boosting domestic tourism include the provision of affordable accommodation and transport to locals in protected areas, raising awareness on the attractions as well as instilling the culture of visiting local attractions among Tanzanians.

Furthermore, diversifying revenue-generating options in WMAs and promotion of income-generating activities that are compatible with nature conservation can also serve as mechanism to cope with the global effects of COVID-19 and other pandemics. The livelihood conditions of local communities could be improved through the adoption of alternative livelihood options that reduce community dependency on tourism for livelihoods. For instance, beekeeping is among the income-generating activities that have a high potential to improve the local livelihoods and is considered compatible with conservation. As a custom, beekeepers need to maintain the natural habitat and reduce the unsustainable use of trees as a source of energy to boost the production of bee products. Many cultures around the globe highly value bee products and if well managed, beekeeping is likely to provide a reliable income for the majority of vulnerable communities that are dependent on tourism.

Moreover, concerted efforts from various stakeholders including the government, Non-Governmental Organizations, and individuals are required to minimize the negative effects of the COVID -19 pandemic in the WMAs. Currently, many WMAs rely entirely on tourism income and partly on donor support to finance conservation activities. There is no specific mechanism that exists for the central government to financially support the WMAs. The right support and funding models would be needed to allow WMAs to sustain their crucial management operations and support local development projects. This includes among other things, restructuring WMA benefit-sharing arrangements to provide the WMA's member villages with more revenue retention by reducing shares to the Government.

6. Conclusion

This study aimed to understand the nature and extent of the COVID-19 pandemic's impacts on the selected WMAs found in Northern Tanzania. In particular, the study looked at the impact of declined tourism revenues on conservation of wildlife resources, local employment as well as the potential trickle-down livelihoods effects on the families of community members. This study revealed that COVID-19 pandemic is having considerable devastating effects on the local livelihoods and WMA revenue collections. The pandemic has affected the incomes of enterprises and individuals due to the closure of tourism businesses and the collapse of the community's small-scale enterprises that are heavily dependent on tourism within the WMAs. Loss of tourism revenue further threatens the conservation of resources and management of WMAs to carry out critical operations for the conservation of wildlife and provide social and economic incentives to the local communities. The current situation thus requires actions to cope with the negative effects of COVID-19 to sustain the functions of the WMAs. Measures put forth include the promotion of domestic tourism, diversification of livelihood options for communities surrounding the WMAs, and concerted efforts by the Government, Non-Government Organizations, and the donor's community to provide funding for the sustainability of the WMAs.

Author details

Rehema Abeli Shoo, Elizabeth Kamili Mtui, Julius Modest Kimaro, Neema Robert Kinabo, Gladys Joseph Lendii and Jafari R. Kideghesho*
College of African Wildlife Management, Mweka, Moshi, Tanzania

*Address all correspondence to: jafari.kideghesho2@mwekawildlife.ac.tz

References

[1] Songorwa AN, Bührs T, Hughey KFD. Community-Based Wildlife Management in Africa: A Critical Assessment of the Literature. Nat Resour J. 2000;40(3):603–43.

[2] Nelson F, Agrawal A. Patronage or Participation? Community-based Natural Resource Management Reform in Sub-Saharan Africa. Dev Change. 2008 Jul;39(4):557–85.

[3] Pailler S, Naidoo R, Burgess ND, Freeman OE, Fisher B. Impacts of Community-Based Natural Resource Management on Wealth, Food Security and Child Health in Tanzania. Ryan SJ, editor. PLOS ONE. 2015 Jul 17;10(7): e0133252.

[4] Adams WM, Hulme D. If community conservation is the answer in Africa, what is the question? Oryx. 2001 Jul;35 (3):193–200.

[5] WWF. Tanzania's wildlife management areas—a 2012 status report. World Wildlife Fund for Nature Tanzania, Dar es salaam.; 2014.

[6] Joyce Lepetu, Richard O. B Makopondo, Michael B.K. Darkoh. Community based natural resource management and tourism partnership in Botswana: which way forward? Botsw Notes Rec. 2008;Vol. 39:113–24.

[7] Stone MT, Nyaupane GP. Protected areas, wildlife-based community tourism and community livelihoods dynamics: spiraling up and down of community capitals. J Sustain Tour. 2018 Feb;26(2):307–24.

[8] IRA. Assessment and Evaluation of the Wildlife Management Areas in Tanzania [Internet]. Institute of Resource Assessment; 2007. Available from: https://www.ucl.ac.uk/pima/docs/reference/15_wma_review.pdf

[9] URT. The Wildlife Policy of Tanzania. United Republic of Tanzania, Dar es Salaam, Tanzania.; 1998.

[10] Kideghesho, JR, Tarimo L. Recent research on CWM in Tanzania – Successes, benefits, and challenges. [Internet]. Wildlife for Communities in Tanzania: Taking stock of governance of wildlife by communities. A Roundtable Meeting 2–3 November 2011. Giraffe Hotel, Dar es Salaam.: TNRF; 2011. Available from: www.tnrf.org › files › CWM_Roundtable_Meeting_Re..

[11] USAID. Tanzania Wildlife Management Areas (WMA) Evaluation. Final Evaluation Report. [Internet]. 2013. Available from: pdf.usaid.gov › pdf_docs › pdacy083.

[12] Wilfred P. Towards Sustainable Wildlife Management Areas in Tanzania. Trop Conserv Sci. 2010;3(1): 103–16.

[13] URT. The Wildlife Policy of Tanzania. United Republic of Tanzania, Dar es Salaam, Tanzania.; 2007.

[14] Alan Rodgers, Lota Melamari, Fred Nelson. Wildlife Conservation in Northern Tanzanian Rangelands. In College of African Wildlife Management Mweka, Tanzania,; 2003.

[15] Fred Nelson. Community-based Tourism in Northern Tanzania: Increasing Opportunities, Escalating Conflicts and an Uncertain Future. In Arusha, Tanzania,; 2003.

[16] Nelson F, Nshala R, Rodgers W A. The Evolution and Reform of Tanzanian Wildlife Management. Conserv Soc. 2007;5:232–61.

[17] Nyamsingwa H.I. Assessment of the Impact of Community Conservation Banks on Conservation and Livelihood in Serengeti District. [MSc

Dissertation,]. University of Dodoma, Tanzania; 2013.

[18] Huang C, Wang Y, Li X, Ren L, Zhao J, Hu Y, et al. Clinical features of patients infected with 2019 novel coronavirus in Wuhan, China. The Lancet. 2020 Feb;395(10223):497–506.

[19] WHO. Director-General's opening remarks at the media briefing on COVID-19 -11 March 2020 [Internet]. 2020a [cited 2020 May 10]. Available from: https://www.who.int/dg/speeches/detail/who-director-general-s-opening-remarks-at-the-media-briefing-on-covid-19

[20] Armstrong F, Capon A, McFarlane R. Coronavirus is a wake-up call: our war with the environment is leading to pandemics. The Conversation, [Internet]. The Conversation. 2020 [cited 2020 Nov 5]. Available from: https://theconversation.com/coronavirus-is-a-wake-up-call-our-war-with-the-environment-is-leading-to-pandemics-135023

[21] Tobias, J., D'Angelo, C. Environmental destruction brought us COVID-19. What it brings next could be far worse [Internet]. 2020 [cited 2020 May 18]. Available from: https://ipbes.net/node/36975. Retrieved on

[22] WWF. The Loss of Nature and Rise of Pandemics: protecting Human and Planetary Health. WWF International, Gland, Switzerland; 2020. 19 p.

[23] Nelson F. The evolution and impacts of community-based ecotourism in northern Tanzania [Internet]. International Institute for Environment and Development 3 Endsleigh Street London WC1H 0DD; 2004 Nov [cited 2020 Sep 25]. Report No.: 131. Available from: https://pubs.iied.org/pdfs/9507IIED.pdf

[24] UNWTO. International Tourist Arrivals Could Fall by 20–30% in 2020 [Internet]. The World Tourism Organization; 2020 [cited 2020 Apr 24]. Available from: https://www.unwto.org/news/international-tourism-arrivals-could-fall-in-2020,

[25] The World Tourism Organization (UNWTO). International tourist numbers could fall 60–80% in 2020 [Internet]. UNWTO. 2020 [cited 2020 Sep 30]. Available from: https://www.unwto.org/news/covid-19-international-tourist-numbers-could-fall-60-80-in-2020

[26] Beekwilder J. The Impact of the Coronavirus Pandemic on the Safari Industry (October Update). [Internet]. SafariBookings. 2020 [cited 2020 Nov 6]. Available from: https://www.safaribookings.com/blog/coronavirus-outbreak

[27] Loibooki M, Hofer H, Campbell KLI, East ML. Bushmeat hunting by communities adjacent to the Serengeti National Park, Tanzania: the importance of livestock ownership and alternative sources of protein and income. Environ Conserv. 2002 Sep;29(3):391–8.

[28] Knapp E, Peace N, Bechtel L. Poachers and Poverty: Assessing Objective and Subjective Measures of Poverty among Illegal Hunters Outside Ruaha National Park, Tanzania. Conserv Soc. 2017;15(1):24.

[29] Duffy R, St John F. Poverty, poaching and trafficking: what are the links? [Internet]. Evidence on Demand; 2013 Aug [cited 2020 Nov 5]. Available from: https://www.gov.uk/dfid-research-outputs/poverty-poaching-and-trafficking-what-are-the-links

[30] Kideghesho JR. The Elephant Poaching Crisis in Tanzania: A Need to Reverse the Trend and the Way Forward. Trop Conserv Sci. 2016 Mar;9(1):369–88.

[31] Hariohay KM, Ranke PS, Fyumagwa RD, Kideghesho JR,

Røskaft E. Drivers of conservation crimes in the Rungwa-Kizigo-Muhesi Game Reserves, Central Tanzania. Glob Ecol Conserv. 2019 Jan;17:e00522.

[32] URT. The 2018 Wildlife Sub-Sector Statistical Bulletin. The Ministry of Natural Resources and Tourism, Dodoma Tanzania.; 2018.

[33] Hsieh H-F, Shannon SE. Three Approaches to Qualitative Content Analysis. Qual Health Res. 2005 Nov;15 (9):1277–88.

[34] Stemler S. An overview of content analysis. Pract Assess Res Eval [Internet]. 2001;7(17). Available from: https://moodle.jku.at/jku/pluginfile.php/27600/mod_resource/content/1/Stemler_2001_An_overview_of_content_analysis.pdf

[35] Holechek J, Valdez R. Wildlife Conservation on the Rangelands of Eastern and Southern Africa: Past, Present, and Future. Rangel Ecol Manag. 2018 Mar;71(2):245–58.

[36] D'udine FAC, Henson DW, Malpas RC. Wildlife law enforcement in Sub-Saharan African protected areas : a review of best practices [Internet]. IUCN, International Union for Conservation of Nature; 2016 [cited 2020 Nov 5]. Available from: https://portals.iucn.org/library/node/46088

[37] Ndosi C. COVID19 TZ Realities — The Economic Impact on Entrepreneurs, Startups, Small and Medium Enterprises in Tanzania [Internet]. The Lauchpad. 2020 [cited 2020 Oct 30]. Available from: https://thelaunchpad.or.tz/2020/04/27/covid19tzrealities-the-economic-impact-on-entrepreneurs-startups-small-and-medium-enterprises-in-tanzania/

[38] Gössling S, Scott D, Hall CM. Pandemics, tourism and global change: a rapid assessment of COVID-19. J Sustain Tour. 2021 Jan 2;29(1):1–20.

[39] Hampson, K, McCabe, J.T, Estes, A. B, Ogutu, J.O, Rentsch, D, Craft, M.E, et al. Living in the greater Serengeti ecosystem: human-wildlife conflict and coexistence. In 2015. p. 607–48.

[40] Nyhus, Philip J, Osofsky, Steve A, Ferraro, Paul, Fischer, H, Madden, Francine. Bearing the costs of human-wildlife conflict: The challenges of compensation schemes. Faculty Scholarship. 15; 2005.

[41] Slocum SL, Backman KF, Robinson KL. Tourism Pathways to Prosperity: Perspectives On the Informal Economy in Tanzania. Tour Anal. 2011 Jan 1;16(1):43–55.

[42] URT. Small and Medium and Enterprises Development Policy. Ministry of Industry and Trade, Dar es Salaam, Tanzania.; 2002.

[43] Mfaume, R. M, Leonard, W. Small Business Entrepreneurship in Dar es Salaam –Tanzania: Exploring Problems and Prospects for Future Development. In Somerset West, South Africa; 2004.

[44] UKEssays. Role of Small and Medium Enterprises in Tanzania [Internet]. 2018. Available from: https://www.ukessays.com/essays/economics/role-of-small-and-medium-enterprises-in-tanzania-economics-essay.php?vref=1

[45] Mzomwe Y, Pelagia M. Capacity of Tanzanian Micro, Small and Medium Enterprises (MMSES) in Tapping the Business Opportunities in the East African Community. Bus Educ J. 2015;1 (1):20.

[46] Ministry of Natural Resources and Tourism (MNRT). Statistical bulleting. 2019.

Chapter 7

A Step Change in Wild Boar Management in Tuscany Region, Central Italy

Paolo Banti, Vito Mazzarone, Luca Mattioli, Marco Ferretti, Andrea Lenuzza, Rocco Lopresti, Marco Zaccaroni and Massimo Taddei

Abstract

In this chapter, reducing the high-density populations of wild boars in an Italian's Tuscany region is addressed as a measure of controlling crop damage and road accidents. The issue is usually tackled from a technical and rarely sociological point of view, making the proposed and implemented solutions less effective. The results presented in these chapter highlight the importance of awareness of the social context when the technical choices are applied. The management of ungulates creates economic interests that oppose changes that shift the economic balance, even when the actions taken are for the benefit of the entire community'. In the previous decades, the wild boar populations have increased considerably in Italy in the Tuscany region. As a consequence of this phenomenon, damage to crops and road accidents has increased. In 2016, the Tuscany region enacted a law to change the management of ungulates by promoting individualism in unsustainable harvest rate areas, allowing shooting wild boar with stalking and selling the meat and maintaining a corporate approach in sustainable harvest rate areas. In three years of enforcing the law, damage to crops and road accidents have decreased significantly and meet supply chain has started. On the other hand, a strong reaction against this Law by wild boar drive hunters emerged. The region is, consequently, faced with an emblematic case where political intervention in future is inevitable in order to mediate between long-term results and short-term consensus.

Keywords: wildlife management, drive hunt, stalking, crop damages, vehicle accidents

1. Introduction

Wild boar (*Sus scrofa*) is among the widest-ranging mammals on the Earth. In Europe, from the 1960s to date, the population has grown dramatically, and its distribution range has expanded [1]. The species has a remarkable ability to adapt to different habitats; this has fostered its spread throughout the European continent, where only three limiting conditions for the habitat requirements have been described: vegetation providing shelter from predators, water for drinking and bathing, and the absence of regular snowfall [2, 3]. The increase in population size has exacerbated crop damages and social conflicts [4–7]. The amount paid by hunters and governments for crop damages caused by wild boar in Europe amounts to millions of Euros [8, 9], and this has increased the effort for preventative methods [10], hunting included [11]. Moreover, the increase of wild boar-vehicle accidents followed by a rise in costs, and people injured should not be underestimated [12]. Some costs can be easily estimated, but others cannot. For example, the cost over the years for seriously injured people's health care is difficult to estimate and usually not considered. European wild ungulates management cope with these challenges under different approaches related to national regulations.

2. Policy and legal governing wildlife and problem animals

In Italy, the management and conservation of wild fauna are mainly organised by public institutions and, to a lesser extent, by private institutions. The current legislation (Law 157/92), transposed the directives of the European Union 79/409/EEC, 85/411/EEC, 91/244/EEC, but in 1992 problems related to the management of ungulates were not particularly felt because the strong demographic expansion took place in the following years. The legislation was inadequate, thus was integrated in 2005 with the national law "248/2005, article 11-quaterdecies, paragraph 5", in which the hunting periods and hours for ungulates were extended. However, the expansion of the ungulate species present in Italy continued. If this was evaluated positively for roe deer and red deer, the same cannot be said for wild boar. The demographic growth of wild boar, favoured by the abandonment of the countryside, by the reduced presence of predators by the illegal releases of subjects coming from other European countries and by foraging, has become the main problem related to agricultural crops' damage and road accidents. The Italian regions organised into provincial or sub-provincial territorial management areas (ATC) have not been able to address the problem effectively with the current legislation. In Tuscany, where traditional wild boar hunting is a cultural heritage, this problem is more evident because ungulate management aimed for decades to increase the abundance of this species for hunting purposes. In this framework, we present the possibility of shifting to a more elastic management model, adapting the hunting periods and hunting techniques to local conflicts with human activities. In 2016 a Regional law (10/2016) brought in Tuscany significant changes in the approach of the ungulate management. This law aimed to reduce, within three years, the agricultural damages caused by wild ungulates, vehicle accidents and strengthen the bush meat supply chain.

Tuscany Region spans an area of 2,298,500 hectares, from the sea to the Appenine mountains, woodland cover the 47% and agricultural areas characterised by vineyard olive grove and cerals cover the 43%. Florence is the biggest city of the region with 800.000 people, the other city Pisa, Arezzo, Siena, Livorno, Grosseto, Lucca, Prato and Massa-Carrara, Pistoia are smaller and with the other urban areas cover the 10% of the territory (**Figure 1**). As in the rest of the country, there is no reliable estimate of the wild boar population, but it is possible to have a rough idea from last year's hunting bag. Wild boars culled from 2015 to 2019 ranged from 70,384 to 96,042 per year (**Table 1**). These data suggest that Tuscany can be considered among the regions with the highest wild boar density in Europe. Ungulate damages continuously increased from 2000 to 2017, and in the last years, the amounts paid to farmers exceeded 2,000,000€ per year (**Table 2**), mainly caused by wild boar to vineyards and cereals. Simultaneously, the economic efforts to prevent crop damages increased, reaching more than 500,000€ per year. A mean of 690 road accidents involving ungulates was recorded every year in Tuscany (2012–2015 average of

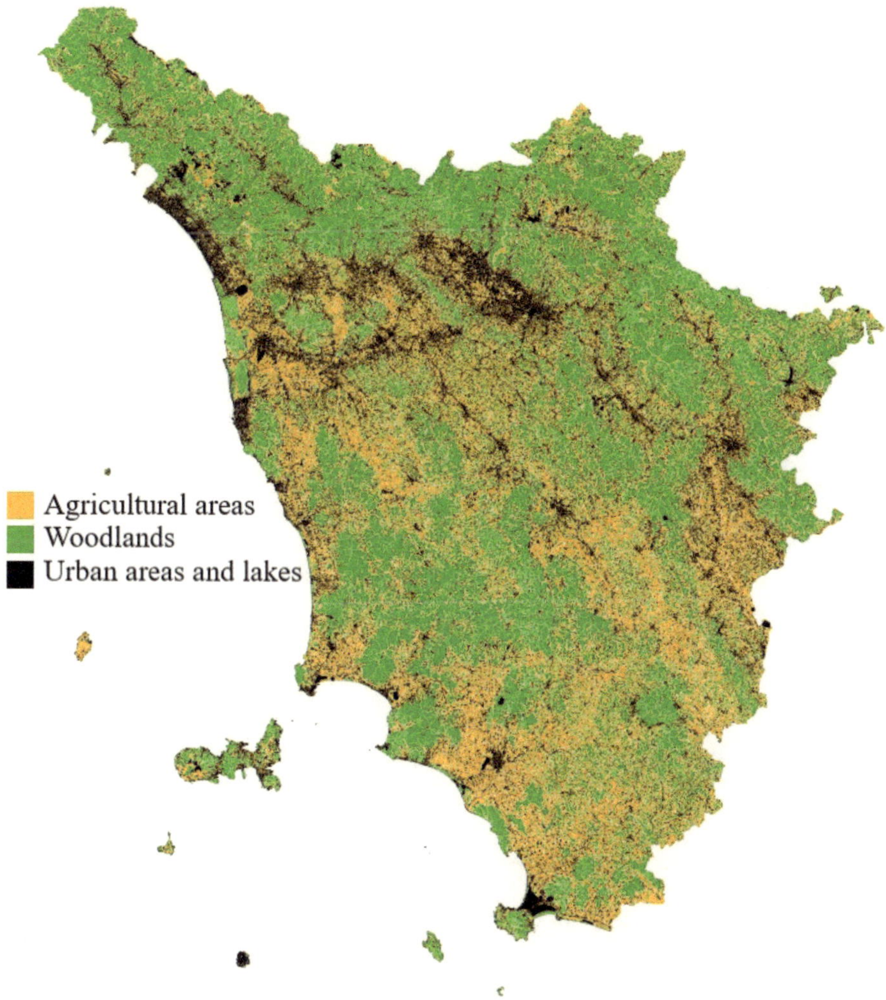

Figure 1.
Study area.

Wild boar hunting techniques	2015	2016	2017	2018	2019
Drive hunt in SHRA	67701	74815	62109	56135	55061
Stalking in UHRA	629	4581	8445	6226	6670
Police officers	10029	9927	13569	10775	5959
Total	79330	96042	88817	76829	70384

Table 1.
Wild boars shot from 2015 to 2019.

claims reported), with consequent material damages, injuries and, in some cases, deaths. In this scenario, the police officers' ungulates culling shifted in recent years from an extraordinary activity, to ordinary and generalised practice, with more than 40,000 culling events per year. The number of ungulates shot or trapped by police officers represents 10% of the hunting bag. The constant decrease of hunters in previous years has resulted in a reduction of the hunting pressure, thus the failure to set reliable goals for the harvest of ungulate populations by using traditional

Species/Years	2010	2011	2012	2013	2014	2015	2016	2017	2018	2019
Roe deer	185.848 €	165.943 €	340.853 €	290.174 €	301.874 €	452.947 €	519.391 €	837.573 €	157.362 €	162.017 €
Red deer	76.506 €	59.871 €	263.291 €	249.185 €	199.296 €	42.156 €	40.435 €	50.951 €	47.799 €	16.986 €
Wild boar	1.049.262 €	1.115.477 €	1.188.767 €	1.032.953 €	1.347.308 €	2.072.198 €	1.792.023 €	2.181.951 €	841.416 €	884.571 €
Fallow deer	46.083 €	51.454 €	59.166 €	82.488 €	73.468 €	67.823 €	80.834 €	122.290 €	20.731 €	20.381 €
Muflon	10 €	40 €	0 €	0 €	12 €	0 €	0 €	0 €	0 €	0 €
Underterminated ungulates	1.085 €	9.164 €	13.435 €	879 €	7.544 €	0 €	0 €	0 €	0 €	12.654 €
Damage caused by ungulates	1.358.784 €	1.401.949 €	1.865.512 €	1.655.679 €	1.929.503 €	2.635.124 €	2.432.683 €	3 192.765 €	1.067.308 €	1.096.609 €
Damage caused by wildlife	1.620.604 €	1.692.474 €	2.112.086 €	2.017.955 €	2.286.166 €	2.929.130 €	2.864.055 €	3.390.665 €	1.114.569 €	1.177.742 €
% Damage wild boar/total	65	66	56	51	60	60	63	64	75	75
% Damage ungulates/total	84	83	88	82	84	83	85	94	96	93

Table 2.
Crop damages divided per species.

techniques. This has prompted two important questions: What should be done when the traditional ungulate management system has proved inadequate to solve the problems described above? What solutions should be included in a framework where the wild boar population is increasing in tandem with growing damage to agriculture and vehicle accidents while economic and human resources are decreasing?

The law enacted in 2016 has attempted to address these challenges through the following four interventions:

1. Ungulate management differentiation between sustainable and unsustainable harvest rate areas.

2. Adopt ordinary hunting activity as an alternative to systematic culling conducted by police officers.

3. Planning ungulate population management in districts.

4. Support the creation of a meat supply chain for wild ungulates.

These interventions are briefly discussed in this chapter.

2.1 Ungulate management differentiation between sustainable and unsustainable harvest rate areas

Sustainable harvest rate areas (SHRA), mainly characterised by woody and bushy areas, were organised with a conservative ungulate population approach (**Figure 2**). On the other hand, unsustainable harvest rate areas (UHRA) aimed to reduce the population substantially. The territory was classified as SHRA or UHRA for each ungulate species based on agricultural damages recorded and potential impact on crops. In Tuscany, almost 50% of the territory is currently UHRA for wild boar and 24% for roe deer (*Capreolus capreolus*). Red deer (Cervus elaphus) has SHRA across the Apennine mountain areas. Fallow deer (*Damadama*) and mouflon (*Ovis aries*) are present in small and localised populations; thus, the management of these species is easier.

In UHRA, the hunting period has been extended for stalking. The harvest rate was set to remove 100% of the abundance estimated during the census, plus the expected increase. The Law sought to reduce ungulate populations, increasing hunting pressure in agricultural and around the urbanised areas. Little or nothing changed in the ungulate management of SHRA, where the target is the conservation of the species.

2.2 Adopt ordinary hunting activity as an alternative to systematic culling conducted by police officers

One of the most important innovations of the Law 10/2016 was the extension to the whole year of the hunting season for wild boar stalking since it had no impact on non-targeted species. It represents the only hunting method on ungulates that can be allowed even in critical periods for crops and other species' biological cycles. The possibility to use police officers in areas and times of hunting ban was, however, provided. On the other hand, involving police officers in these activities means expensive and complicated procedures (request of the farmer, application of proactive measures, single authorisation act, and coordination of the police officers). For these reasons, the Law aimed to shift from an expensive and extraordinary approach to a profitable and ordinary activity.

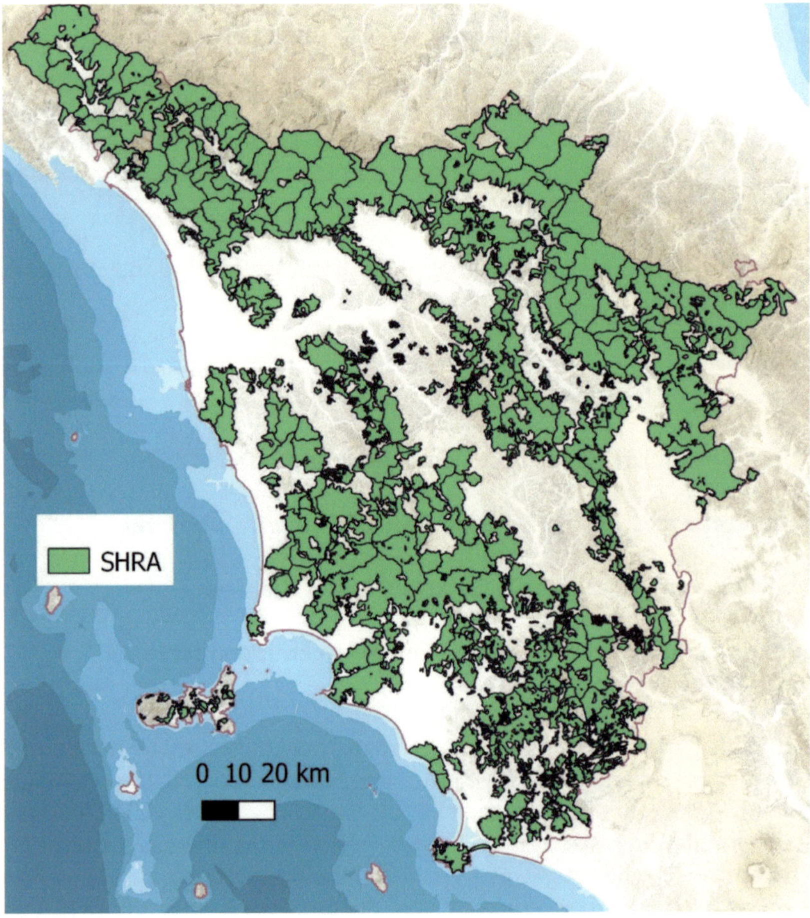

Figure 2.
Wild boar sustainable harvest rate areas.

2.3 Planning ungulate's population management in districts

The Law introduced the concept of district management, big enough to include private and public hunting sub-districts and protected areas. The aim was to overcome the past fragmentation of competencies, standardising census techniques, harvest rate, and protected areas management. Thus, Tuscany has been divided into 800 Ungulate sub-districts coded through an App (Toscaccia) where wildlife technicians can add the census data, hunting bag, and a free access cartographic portal, Geoscope (http://www502.regione.toscana.it/geoscopio/cacciapesca.html).

2.4 Support the creation of a supply chain for wild ungulates meat

Ungulates meat supply chain represents a strategic topic in wildlife management. In UHRA, permits were fixed cheap to increase the hunting pressure, given the possibility to sell the hunted meat, promoting an economical chain among hunters, farmers and meat retailers (game handling establishments, butchers, meat chains and dealers) and consumers. The aim was to transform the "ungulate problem" through rational hunting exploitation into managing a renewable economic resource.

Here we reported three years of this management strategy focusing on wild boar because it was the main problem from an economic and sociological perspective.

In Tuscany, the wild boar drive hunt is the traditional hunting method. Dozens of people organised in teams manage the territory, in particular the SHRA. A few years ago, drive hunt was the only method allowed for wild boar, and teams managed the entire population; thus, the feeling of ownership of wild boars living in those territories was solid. This activity is more than a hunting technique; it represents a recreational activity that involves hundreds of people. In the rural context, the village festival organised by wild boar hunters is one of the most important social events. On the other hand, the wild boar was the main cause of road accidents and crop damage with a growing trend. Thus, it is comprehensible (but not acceptable) that a new management strategy aimed to eradicate the wild boar population in the UHRA caused a social conflict.

The eradicative approach of wild boar populations in UHRA with the stalking method throughout the year is the most novel aspect introduced in 2016 because it increased the hunting pressure and the competition among hunters. UHRA were organised to favour individualism; no limits for the number of people in the districts, no necessity to be accepted in the UHRA from other hunters, no assignment of the number of animals to shot, and no assignment of a hunting area in the UHRA. These rules favour individualism and strongly reduce social control over the hunters. Stalking was in addition to the drive hunt, but the last one is allowed only in SHRA for three months in winter. Although stalking was also allowed in SHRA, it has never been applied due to conflicts with drive hunters that traditionally manage the SHRA.

From 2016 to 2019, almost 26,000 wild boars were culled by stalking in UHRA. In the same period, 248,120 wild boars were culled in SHRA with drive hunt, which means that the impact of stalking on wild boar population is much lower than drive hunt, but stalking was applied on agricultural areas, where human conflicts emerged. Stalking of wild boar showed a peak of culling from April to September (**Figure 3**). However, the hunting bag structure did not represent the population's demographic structure, showing a prevalence of adult males culled (**Figure 4**). Cautiousness is needed to analyse these data because some hunters avoid shooting females when pregnant or with piglets.

Stalking had a more significant impact in spring and summer when other hunting methods were not allowed, mainly when crop damages were more significant. Population control by police officers, usually exploited with drive hunters, was previously carried out in the autumn and winter seasons. However, from 2016 to

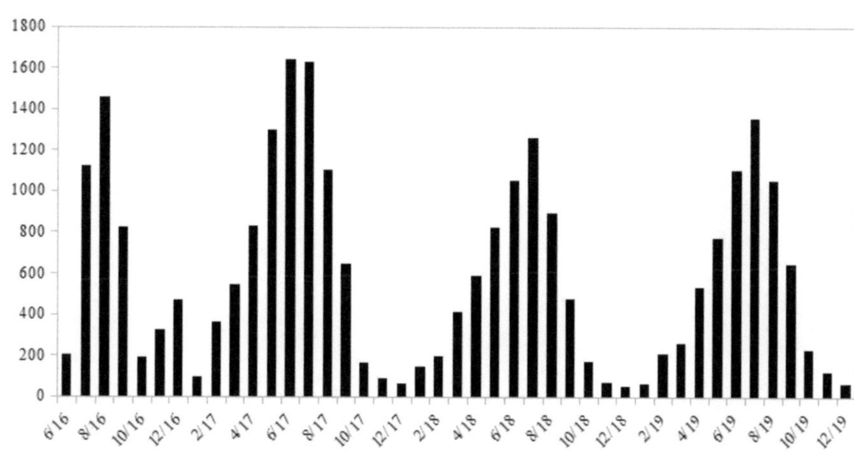

Figure 3.
Wild boars shot by stalking monthly from June 2016 to December 2019.

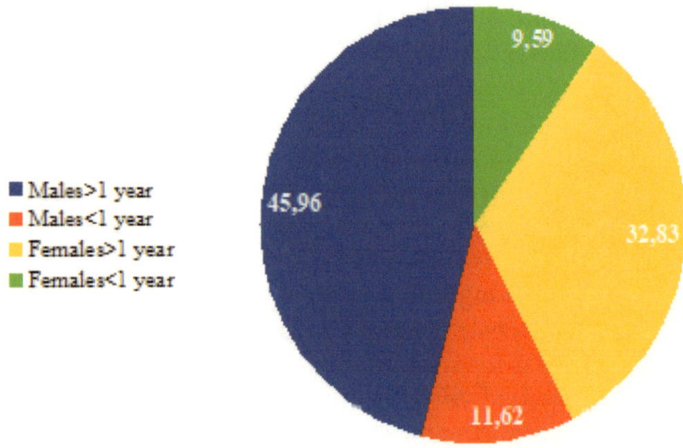

Figure 4.
Demographic structure of wild boars shot from June 2016 to December 2019.

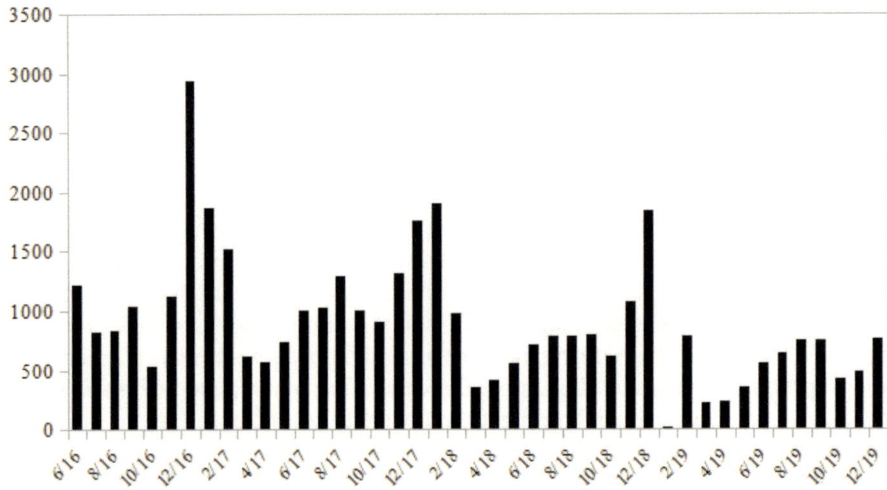

Figure 5.
Wild boars shot under police control monthly from June 2016 to December 2019.

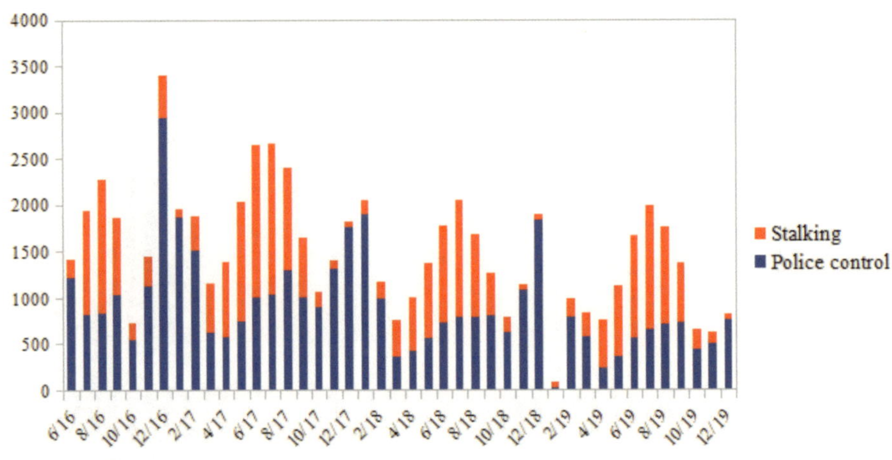

Figure 6.
Wild boars shot by stalking and under police control from June 2016–December 2019.

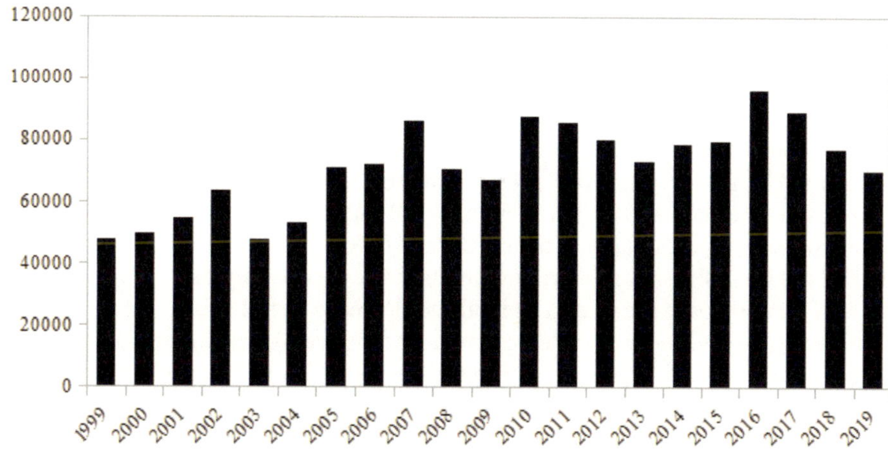

Figure 7.
Wild boars shot with drive hunt, 1999–2019.

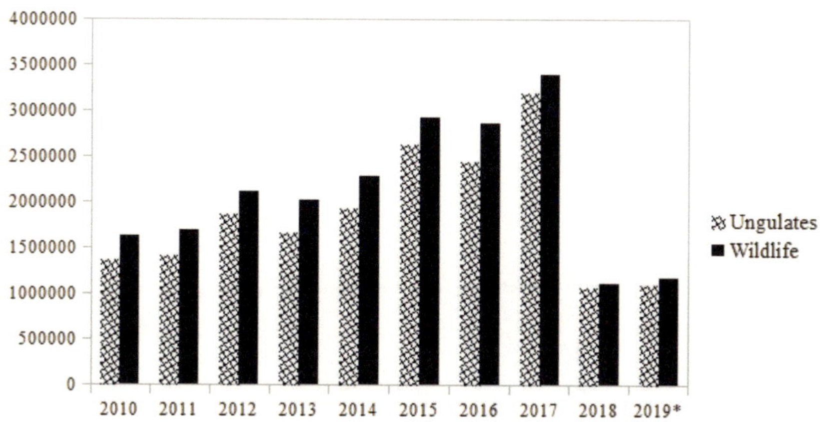

Figure 8.
Wildlife damages to agriculture in €.

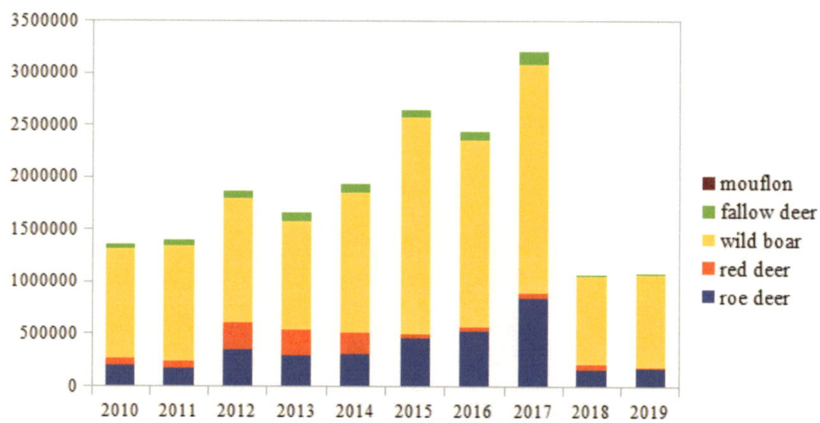

Figure 9.
Damages to agriculture caused by ungulates and wildlife in Tuscany in €.

2019, they became better distributed during the year (**Figure 5**); this means that a significant change of strategy has affected the overall approach to hunting and control, which resulted in complementary during the year (**Figure 6**).

Despite attempts to hinder the start of stalking, many hunters approached this method with great interest. In three years, almost 15,000 boar stalkers were trained in Tuscany. Thanks to the hunting pressure in UHRA, a virtuous mechanism started with an indirect effect on the management of SHRA, where hunting pressure of drive hunters increased, particularly at the border of UHRA. In fact, it was indicated that 2016 had the most significant number of culls in the past twenty years (**Figure 7**). Then, the hunting bags recorded a generalised decrease of culled boars until 2019.

Ungulates caused most of the crop damages (**Figure 8**), and among ungulates, wild boar was the leading problem animal (**Figure 9**). From 2016 to 2019, crop damages and vehicle accidents decreased significantly (**Figures 7** and **10**). The meat supply chain started slowly for the reasons described below because the market prefers animals shot through stalking. In three years, seventeen structures were built where hunters could deliver wild ungulates; fourteen managed by hunting districts, one by a Park Authority and two by the private sector. The available amount of wild boars culled for the supply chain is still below 10% of the wild boar shotdue to logistical/structural problems and killing methods because hunted ungulates are not always suitable for the market (**Figure 11**).

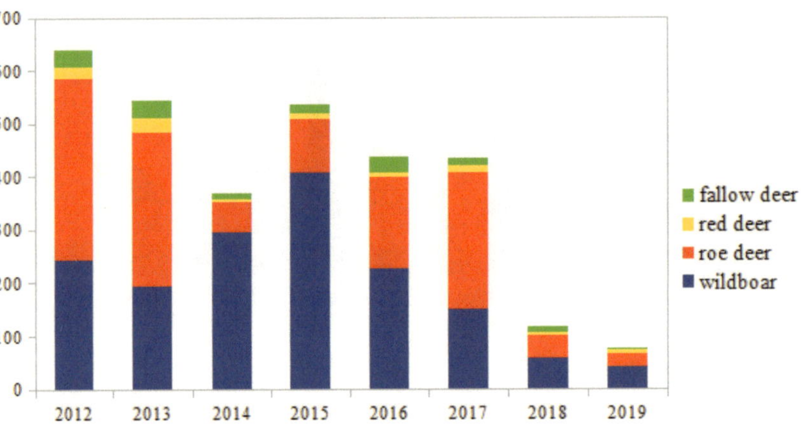

Figure 10.
Traffic accidents involving ungulates.

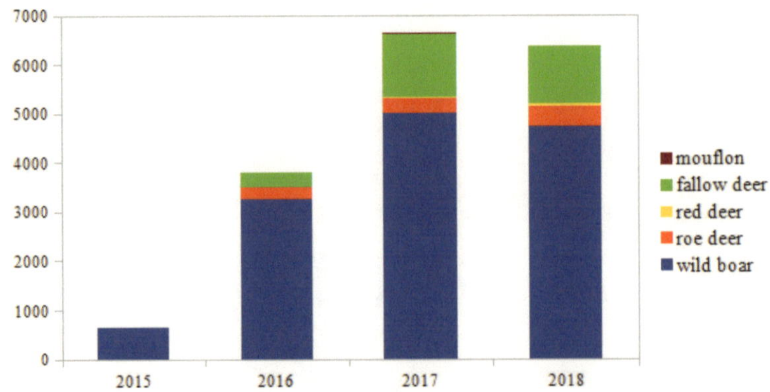

Figure 11.
Ungulate meat supply chain.

3. Conclusion and policy recommendation

As expected, three years were enough to obtain satisfactory results, but social conflicts raised by this law were unexpected. Many economic interests are involved in ungulate management, mostly legal and legit, others illegal, as the meat's black market. Ungulate management, and in particular wild boar management described above, is a strategy that can be replicated in other Italian regions. However, in other countries, its use can be limited by national regulations. Nevertheless, the thread that binds the management of ungulates in different countries, regardless of the regulations, is that to reduce densities, it is necessary to leverage individualism, the possibility of selling culled animals and increasing competition between hunters. From a social point of view, this could be hard to achieve. We often cope with problems caused by ungulates, talking about technical aspects, when most of the management failures are related to difficulties to change hunting traditions based on a conservative approach.

Therefore, it is essential to analyse the socio-economic context on which it is intended to legislate to obtain effective results in the management of ungulates. In Europe, creating a meat supply chain requires more defined and binding legislation capable of overcoming local resistance and promoting a supply chain in which stakeholders will be ready to invest. Politicians, who are sensitive to people's opinions, seek to mediate between what is right and public consensus in the attempt to reach a compromise.

Acknowledgements

Thanks to the colleagues of the Regional Territorial Offices, to the members of the Ungulates Working Group (WG), to the Ambiti Territoriali di Caccia, to the private wildlife institutes and to all the technicians who collaborate for the App "Toscaccia". The data of the meat supply chain were kindly provided by Dr. Alessio Capecci, Tuscany Region.

Author details

Paolo Banti[1], Vito Mazzarone[1], Luca Mattioli[1], Marco Ferretti[1], Andrea Lenuzza[1], Rocco Lopresti[1], Marco Zaccaroni[2*] and Massimo Taddei[1]

1 Wildlife Management Office, Tuscany, Italy

2 Department of Biology, University of Florence, Florence, Italy

*Address all correspondence to: marco.zaccaroni@unifi.it

References

[1] Massei, G., Kindberg, J., Licoppe, A., Gačić, D., Šprem, N., Kamler, J., Baubet, E., Hohmann, U., Monaco, A., Ozoliņš, J., Cellina, S., Podgórski, T., Fonseca, C., Markov, N., Pokorny, B., Rosell, C., and Náhlik, A. (2014). Wild boar populations up, numbers of hunters down? A review of trends and implications for Europe. Pest Management Science 71:492-500.

[2] Tack, J. (2018). Wild Boar (*Sus scrofa*) Populations in Europe: A Scientific Review of Population Trends and Implications for Management. Brussels: European Landowners' Organization. 56 pp.

[3] Marsan, A., and Mattioli, S. (2013). Il Cinghiale (in Italian). Il Piviere (collana Fauna selvatica. Biologia e gestione).

[4] Schley, L., Dufrêne, M., Krier, A., and Frantz, A. C. (2008). Patterns of crop damage by wild boar (*Sus scrofa*) in Luxembourg over a 10-year period. European Journal of Wildlife Research 54:589-599.

[5] Amici, A., Serrani, F., Rossi, C. M., and Primi, R. (2012). Increase in crop damage caused by wild boar (*Sus scrofa* L.): The "refuge effect". Agronomy for Sustainable Development 32:683-692.

[6] Lombardini, M., Meriggi, A., and Fozzi, A. Factors influencing wild boar damage to agricultural crops in Sardinia (Italy). Current Zoology 63(5):507-514.

[7] Valente, A. M., Acevedo, P., Figueiredo, A. M., Fonseca, C., and Torres, R. T. (2020). Overabundant wild ungulate populations in Europe: Management with consideration of socio-ecological consequences. Mammal Review 50(4):353-366.

[8] Maillard, D., Gaillard, J. M., Hewison, M., Ballon, P., Duncan, P., Loison, A., et al. (2010). Ungulates and their management in France. In: European Ungulates and their Management in the 21st Century. eds. Apollonio, M., Andersen, R., and Putman, R. Cambridge, UK: Cambridge University Press. pp. 441-474.

[9] Gren, I. M., Andersson, H., Mensah, J., and Pettersson, T. (2020). Cost of wild boar to farmers in Sweden. European Review of Agricultural Economics 47(1):226-246.

[10] Schlageter, A., and Haag-Wackernagel, D. (2012). Evaluation of an odor repellent for protecting on crops from wild boar damage. Journal of Pest Science 85:209-215.

[11] Geisser, H., and Reyer, H. U. (2004). Efficiency of hunting, feeding and fencing to reduce crop damage by wild boars. Journal of Wildlife Management 68:939-946.

[12] Jägerbrand, A. K., and Gren, I. M. (2018). Consequences of increases in wild boar-vehicle accidents 2003-2016 in Sweden on personal injuries and costs. Safety 4(4):53.

Chapter 8

Managing Invasive Alien Species by the European Union: Lessons Learnt

Ludwig Krämer

Abstract

The contribution concentrates on the fight against invasive alien species within the European Union (EU), which groups 27 States. In 2014, the EU adopted a regulation to identify and manage invasive alien species. This regulation and its monitoring are discussed in detail, in order to see, what lessons can be learnt from the cooperation and concertation of the different states.

Keywords: invasive alien species, European Union, regulation 1143/2014, management of IAS

1. Introduction

The qualification of some wildlife species as "invasive alien species" [1] is a man-made qualification and thus necessarily arbitrary. Indeed, how species expand in the wild should not, one is inclined to think, be determined by humans, but should be left to the natural evolution of biological diversity. However, a closer look into the problem shows that species which entered new ecosystems could have very negative impacts on their new environment. Examples are the Nile perch (lates niloticum) which was introduced into Lake Victoria in the 19th century and caused the extinction of some 200 indigenous fish species; the coulerpa seaweed (Coulerpa taxifolia) invaded the Mediterranean and severely damaged the indigenous aquatic flora and fauna; the introduction of the Polynesian rat into Easter Island is thought to have contributed to the deforestation of that island, with severe consequences for the human populations, etc.

In view of the potential problems caused by invasive species, measures were taken at national and later at international level to stop the further expansion of species outside their natural range. At international level, the Convention on Biological Diversity of 1992 (CBD) asked the Contracting Parties to take measures, in order to "prevent the introduction of, control or eradication of those alien species which threaten ecosystems, habitats or species" [2]. The Aichi Targets of 2010, adopted under the auspices of the CBD, formulated in a similar way [3].

The European Union (EU), which had adhered to the CBD in 1993, adopted a strategy on biodiversity in 2011, which took over the Aichi Target 9 almost word by word [4]. Subsequently, EU Regulation 1143/2014 on invasive alien species was adopted which introduced, for the first time, EU-wide provisions on the fight against invasive alien species (IAS) [5].

More than six years after the adoption of this Regulation it might be time to discuss its merits and weaknesses and examine, whether amendments of the legal provisions at EU level are appropriate. This contribution will thus limit itself on questions of EU law and legal policy with regard to the management of IAS. In a first part, the genesis of the Regulation and the subsequent implementation measures will be described, together with the implementation measures which were adopted by the EU Member States. This is followed by the a presentation of the implementation of the Regulation by Member States and lessons learnt from this process. A short final remark ends the contribution.

2. The elaboration of regulation 1143/2014

The taking of EU measures on IAS was first decided in the Sixth EU Environment Action Programme of 2002 [6]. Following a number of other Commission communications and external studies on IAS [7] and in particular a very detailed impact assessment of possible EU initiatives on IAS [8], the Commission presented, in 2013, a proposal for a regulation on IAS [9]. The European Parliament [10] and the European Economic and Social Committee [11] only suggested minor amendments, which were easily acceptable to the Council, so that the Regulation could be adopted within thirteen months after its proposal, an unusually short period of time for an EU legislative text.

Regulation 1143/2014 was supplemented by some technical provisions [12] and by the establishment of a list of IAS of Union concern [13]; this list was, up to mid-2020, two times updated [14].

The Commission's impact assessment had identified two main reasons for legislating at EU level: first, the ecological problem: IAS caused considerable economic, social and environmental damage. As the introduction of IAS into the EU had increased, between 1970 and 2007, by 76 per cent [15] and was very likely to further increase, due in particular to increased trade and mobility and the impact of climate change, the costs of combating IAS and reducing damage were also likely to increase; Union action thus became necessary. The second reason which was identified by the impact assessment was the fact that "the policy failure caused by a very fragmented and incoherent policy set up at EU and national levels which allowed the ecological problem to worsen" [16]. Apparently, these reasons were accepted by the EU legislature, the European Parliament and the Council.

3. The content of the IAS legislation

The Regulation 1143/2014, structured in six chapters [17], pursued three main objectives: the coordination between the EU and its Member States, the prevention of IAS to enter or spread within the EU, and the prioritisation and management of measures.

3.1 Coordination

The coordination objective of Regulation 1143/2014 was first of all reached by the very fact of establishing EU-wide legislation on IAS. Indeed, an EU regulation is of general application, binding in its entirety and directly applicable in all Member States [18]. According to the well-established principle of EU law that EU law prevails over national law, all national legislation which contradicted the provisions of Regulation 1143/2014, became inapplicable. This also applied to national legislation, which was adopted after the entry into force of Regulation 1143/2014.

A second important element of coordination was the establishment of a common terminology and language [19]. The Regulation established 17 common definitions, among them the terms "IAS", "IAS of Union concern", "IAS of Member State concern", but also terms such as "introduction", "eradication" or "widely spread" [20]. In this way, it influenced national, regional and local regulation, administrative practice and also scientific research on IAS.

Article 22 of the Regulation laid down the general obligation of Member States to coordinate their activities in combating IAS, specifying the conditions which would make cooperation and coordination particularly desirable [21]; they may invite the EU Commission to facilitate the cooperation [22]. Furthermore, Member States should make efforts to ensure coordination and cooperation with third countries, where this is appropriate.

These coordination and cooperation objectives were further specified in different other provisions. For example, when a Member States takes emergency measures, in order to react to a new appearance of an IAS, it is obliged to inform the Commission and the other Member States [23]; the implication is that this might lead to joint efforts of different Member States. Problems of only regional concern and IAS which are of concern to one Member State shall again be addressed by coordinated and cooperative action [24]. Action plans for addressing the pathways of introduction and spread of IAS shall preferably be coordinated at the appropriate regional level [25]. Also management measures for IAS that are already widely spread, shall be notified, where appropriate to other Member States which might be concerned. The Regulation strongly favoured coordinated management measures in such cases [26].

Coordination and cooperation was, moreover, favoured by the establishment of a number of bodies which blossom under the Regulation and which are chaired by the Commission. An IAS Committee, consisting or representatives of the 27 Member States, assists the Commission in all questions, such as the establishment of IAS lists or amendments of the Regulation [27]. A scientific forum assembles scientists appointed by the Member States to assist the Commission in scientific questions [28]. An IAS expert group advises at the initiative of the Commission; its compositions is largely similar to that of the Committee, though the experts do not represent their State of provenance. Finally, there is a body which assembles about thirty members of stakeholder groups and of public authorities [29].

It is obvious that the numerous meetings of these bodies and other contacts between their members and with the Commission, and the "soft" invitation to cooperate in the different provisions that were mentioned, led to increased exchange and transfer of know-how, cooperation and coordination among the public authorities of the 27 Member States, scientists, NGOs and other stakeholders within the EU, which is further promoted by the obligation to regularly report on the application and enforcement of the Regulation [30], the availability of EU funding to finance pilot and other projects, local eradication or containment methods [31], risk assessments [32], joint scientific research and publications [33], the joint collection of data on IAS [34] or the genesis of other bodies which specifically address IAS problems. All these activities contribute to the joint venture of fighting IAS within the EU, including in those regions, which, in the past, paid less attention to it.

3.2 The species regulated

From the great number of invasive alien species, Regulation 1143/2014 only regulated those which were considered to be "of Union concern", i.e. species whose adverse impact "required concerted action at Union level" (Article 3). It is true that also IAS of Member State concern were referred to in the Regulation; however, as Member States were anyway allowed to introduce or maintain more stringent legal requirements at national level than those which were laid down in Regulation 1143/2014 [35], this provision did not have an additional legal value [36].

The EU list of IAS is the core element of the Regulation. By mid-2020, it contained 66 IAS. While the original Commission proposal, to limit the number of IAS of Union concern to 50 species, was not retained, it may be expected that the "comitology" procedure [37] will have as a consequence that the list does not become too long, as this would increase the workload for national authorities; and the Member States are not prevented from becoming active in their territory and also to coordinate with neighbouring countries with regard to species that are not on the EU list, but are of national concern.

In order to be inserted in the list of IAS of Union concern, a species must have undergone a risk assessment and be likely to cause significant damage; this risk assessment includes the implementation costs, the costs of inaction, the cost-effectiveness and socio-economic aspects [38]. It is not necessary that a species is already present in the Union in order to be included in the EU list. The Commission has to propose to the Committee, which was set up under Article 27 of Regulation 1143/2014, the inclusion of a species in the list of Union concern. The Committee decides with qualified majority on the proposal; the final decision is taken by the Commission, which may, though, not go against the opinion of the Committee [39].

For listed species, a number of restrictions concerning the intentional introduction into the EU apply, such as an import, trade or transport ban, the prohibition to keep or release the species or to let it breed. Also the unintentional introduction or spread of the listed species is to be prohibited by Member States [40]. Member States may exceptionally grant permits for research, ex-situ conservation in contained holdings or, for reasons of compelling public interests (including economic interests) and subject to authorisation by the EU Commission, other uses [41]. Restriction measures may also be taken by a Member State regarding a species that is not listed, but should be listed. In such a case, an EU-procedure is initiated, where the Commission decides, whether or not to propose the inclusion of the species in the EU list [42].

Furthermore, the inclusion of an IAS in the list of Union concern obliged the Member States to:

- elaborate action plans on the pathways for unintentional introductions of IAS used and inform the Commission (Article 13);
- instal a surveillance system, in order to prevent the spreading of the IAS (Article 14);
- introduce a border control system (Article 15);
- inform the Commission and the other Member States of the introduction or presence of an IAS, whose presence was previously unknown (Article 16);
- provide for the rapid eradication of the IAS at an early stage of detection, unless one of the exceptions of Article 18 applies and the derogation from the obligation is accepted by the Commission (Articles 17 and 18).

The Regulation provided for limited obligations with regard to IAS which are already widely spread. Member States shall aim to minimise the damage caused by such IAS and provide for their eradication, control or containment. Their management methods shall be the subject of strict economic (cost) considerations. Member States shall also try to restore damaged ecosystems. It is of particular relevance that it is each Member State which decides, whether an IAS is widely spread and, following, what management measures it will take. The general cooperation and coordination obligations of Article 22 also applies to the management measures concerning the widely spread IAS.

3.3 Transparency and public participation

The transparency requirements of Regulation 1143/2014 and in particular its monitoring by the Commission are not optimal. It is not known, what proposals the Commission submits to the Committee which was set up under Article 27 and what arguments are used by which Committee member to accept or refuse the proposal for a new IAS to be inserted in the list of Union concern. The names of the members of the scientific forum, established under Article 28, are not made public [43]. Minutes of committee meetings are not published. And the composition of the stakeholder body which groups individual experts, NGOs, economic operators and public authorities and which advises the Commission, is not either known.

The EU Member States had to inform the Commission of their national legislation concerning penalties, in particular for not complying with the restrictions under Article 7 of the Regulation. The Commission does not publicly make available the national legislation. It only mentioned, at a hidden place, that some Member States had not aligned their legislation to the requirements of the Regulation [44].

Member States were also obliged to inform the Commission and other Member States concerned of the detection of an IAS that was previously unknown in their territory; however, some Member States had not notified the Commission of the eradication measures, which they were obliged to take with regard to such early detected species (Article 17) [45]. If and what measures the Commission took to enforce these and all other existing legal obligations, is not known.

The Member States' reports concerning the implementation of Article 24 [46] - all Member States reported, with the exception of Portugal - are available on the internet [47], the same as information on recent detections of IAS species (Article 16) [48].

Article 26 of the Regulation obliged Member States to let the public participate in the elaboration of action plans for the unintentional introduction of IAS (Article13) and the management measures undertaken according to Article19. No obligation exists for the Commission to let the public participate in the elaboration of measures to insert new species of Union concern in the common list.

4. The results

4.1 The Union list of IAS

The most important result of Regulation 1143/2014 is the establishment of a list of IAS of Union concern. Before the establishment of that list, only some Member States had national lists of IAS and these list were in no way aligned, concerted or otherwise adapted to specific biogeographical regions. The Union list obliges Member States now to report on management measures to prevent the introduction and spread of each of the listed IAS and to report on the results of such measures. While the Regulation allows Member States to abstain from taking management measures with the argument that a specific IAS is widely spread and management measures would be too costly, the Member States are under a certain control by their citizens and the scientific community, which might contest or correct such reasoning.

A closer look at the national reports on the implementation of the Regulation - the Commission will report on its implementation by mid-2021 at the earliest - revels that Member States did not always pay great attention to answer the questions, which the Commission had asked in Regulation 2017/1454 [49]. Also, the Commission itself doubted in several cases, whether the poor data availability had not led Member States to report that an IAS was not present in their territory [50]. In particular the impact of management measures on non-targeted species was hardly ever assessed and commented. Moreover, the question whether the management measure had led to an eradication, a population decrease or increase, or whether the IAS population remained stable or the population trend was unclear, is rather general and allowed answers ("unclear"), which were not always based on thorough assessment.

The list of, until now, 66 IAS of Union concern shall be regularly updated, but it will certainly not be possible - and perhaps not even desirable - to increase the number of IAS of Union concern to 900, a figure which was mentioned in scientific publications as necessary [51]. Such a scientific request overlooks the fact that Regulation 1143/2014 explicitly included considerations of cost-effectiveness and the capacity of the national and regional authorities to take effective management measures to combat IAS. It should not be forgotten that no Member State is prevented from taking measures also with regard to IAS which are not on the EU list.

4.2 The early detection of IAS

The prevention of the introduction of IAS into the EU environment is another important objective of the Regulation. Between 2016 and June 2020, Member States notified to the Commission under Article 16 the early detection of - without UK data - 84 IAS of Union concern, concerning 26 different species [52]. Most notifications came from Germany, followed by Netherlands and Ireland [53]. The prsence of IAS Asian hornet (Vespa velutina nigrithorax), Ruddy duck (*Oxyura jamaicensis*) and Munitacus deer (*Muntiacus reevesi*) was most frequently notified [54]. In 34 cases, the Member States indicated that the IAS had been eradicated; in 36 cases the eradication was ongoing and in 14 cases, the IAS was not eradicated. No data are known on the result of border controls or other means to stop the intentional introduction of IAS into the EU environment.

4.3 Widely spread species

The results with regard to IAS which are widely spread, are much less clear. As mentioned, Member States decide on the cost-effectiveness of measures and thus, whether measures should be taken at all. They do not have to explain their decisions. This leads to the situation that for example, Greece, Cyprus, Romania and Bulgaria reported that between 2015 and 2018, they did not undertake one single management action to eradicate, control or contain IAS in their respective territories [55].

Member States had to report, under Article 24 of the Regulation, on the 49 IAS of the EU list and its first updating. They reported on the presence of, overall, 43 IAS [56]. The eleven IAS which were present in the greatest number of Member States were seven animals and four plants, namely Common milkwood (*Asclepias syriaca*) with 23 notifications; Himalayan balsam, 22, (*Impatiens glandulifera*); Muskrat, 22, (Ondatra zibethicum); Slider turtle, 22, (*Trachemys scripta*); Signal crayfish, 21, (*Pacifastacus leniusculus*); Giant hogweed, 21, (*Heracleum mantegazzianum*); Nuttall's waterweed, 19, (*Elodea nuttallii*); Egyptian goose, 17, (Alopochen aegypiaticus); Spiny-cheek crayfish, 17, (Orconectus limosus); Stone moroko, 16, (*Pseudorasbora parva*); and Chinese mitten crab, 16, (Erocheir sinensis).

The following table indicates details of the national reports. It must be stressed that the results of the different measures (columns 4 to 8) are not comparable. Indeed, some Member States, such as Sweden or France, reported on measures with regard to individual populations. Other Member States, such as Slovenia or Netherlands, resumed the different measures in one single national figure. For example, Slovenia reported that a specific IAS had been eradicated in 16 places and that eradication was ongoing in seven more places; yet, the report only signalled one "decreasing number" (**Table 1**).

Member state	Number of IAS present	IAS subject of measures	Result: eradicated	Result: increasing	Result: stable	Result: decreasing	Result: uncertain
Belgium	31	23	1	6	3	2	3
Bulgaria	12	—	—	—	—	—	—
Cyprus	3	—	—	—	—	—	—
Czechia	14	1	—	—	—	1	—
Croatia	18	1	—	—	—	1	—
Denmark	14	2	—	—	—	1	1
Estonia	12	8	—	1	1	3	3
Finland	10	8	2	1	1	2	2
France	34	20	5	9	11	30	23
Germany	26	18	5	15	5	6	51
Hungary	26	8	—	3	3	2	—
Ireland	14	6	1	1	—	3	3
Italy	31	21	1	7	10	17	44
Lithuania	9	5	—	—	—	—	5
Luxemburg	10	6	—	1	—	7	1
Malta	5	2	—	1	1	1	—
Netherlands	29	23	1	1	4	3	16
Poland	16	5	—	—	—	3	8
Austria	22	12	—	—	—	—	12
Romania	13	—	—	—	—	—	—
Slovakia	15	6	—	—	—	—	30
Slovenia	12	10	—	1	1	3	7
Spain	28	19	1	9	6	1	3
Sweden	12	10	20	—	3	15	44
Total			37	56	49	101	256

Table 1.
Invasive alien species in EU member states, according to the member states' reports to the European Commission.

5. Management lessons learnt

The adoption of Regulation 1143/2014 and in particular of a common list of IAS of Union concern undoubtedly increased the active fight of Member States against IAS, also, because only a minority had national IAS lists [57]. The obligation to notify the Commission of early detected IAS and of the measures taken to eradicate them apparently stimulated national authorities to take active measures to prevent the spreading of such IAS. This may be evidenced by the numerous measures against the Asian hornet, the Ruddy duck or the Coypus (*Myocastor coypus*) which rank high on the notification list under Article 16 of the Regulation [58], but are not particularly far spread in the EU.

Cooperation and concertation of Member States under Regulation 1143/2014 has its limits, though. A species which is not spread in many Member States, may be of very high relevance in individual countries and require action at that level; examples are the Red swamp crayfish (*Procambarus clarkii*) in Spain and the Sosnowsky's hogweed (Heracleum sosnowskyi) in Poland [59]. Hopes to improve the cooperation among Member States should therefore not be too high: even within individual Member States, cooperation is not always perfect; this applies in particular, but not only, to regionalised countries, such as Belgium, Germany, France. For this reason, it would only in exceptional cases make sense to agree EU-wide concerted actions regarding specific species, all the more as only seven Member States host more than half of IAS of Union concern [60]. More sense would be regional cooperation in appropriate cases, for example to prevent the spread of the Raccoon (Procyon lutor).

These findings contradict the reason for listing an IAS on the Union list, because a species of "Union concern" is defined as a species which "requires concerted action at Union level". The national reports do not give the impression that any such concerted action has taken place until now, though it might be too early to draw final conclusions at this stage: concertation is a process and Member States might have to get accustomed to cooperate beyond the national borders. However, it seems unlikely that without strong EU Commission initiatives in this regard - including the (co-)financing of eradication measures - concerted actions by several - not to talk of all - Member States will blossom.

The differentiation between eradication, control and containment for widely spread IAS did not show significant results. When action was taken, this was mostly done in order to eradicate a species, though success was limited, as evidenced by the small number of eradication successes and the great number of uncertain results in the table above.

The action programmes on pathways for IAS introduction (Article 13) were not referred to in the national reports, as they had to be established only by mid-2019. They will thus not be commented in this contribution.

The reports on the cost of the national measures often give the impression that the Member States do not know themselves the amount of cost of the measures, also because such costs form part of the normal work of the responsible staff at local, regional or national level, and no specific, ear-marked sums were made available to fight IAS. It does not seem possible to draw convincing conclusions on the amount of money which was spent to fight IAS.

The involvement of the public was insufficient. It is not clear, whether the authors of Regulation 1143/2014 had in mind that specific local, regional or national projects of the type of LIFE-projects would be decided to fight this or that IAS, and specific sums would be made available. In such a case, public participation is useful and may bring added value. However, the national implementation reports normally show that most countries did not make specific arrangements to fight IAS in general or a specific species, but that "business as usual" continued. The reports thus indicated only in general terms, how the public was informed of measures, plans or projects in the fight against IAS. Information of the public requires an information at local or regional level, where measures against IAS are taken, in order to gain the support of the population. This also requires that the language on IAS regulations, plans and measures gets away from exclusively using the Latin name of the species and refers to the species' name in the local language. The same is true for public participation: there is need to show, what damage is caused by the IAS and what citizens can do to improve the situation, by actively assisting in the early detection of new IAS and in the fight against widely spread IAS. As on all this, Regulation 1143/2014 was too general, the reaction of Member States also remained general.

Apparently, the Commission intends to regularly update - increase - the number of IAS of Union concern. To the extent that this will increase the workload of the local etc. authorities which deal with IAS problems, it is not likely to lead to better results in fighting IAS [61]. It might be more promising to seek concertation of the different Member States concerned, in order to eradicate for example, the five most invasive species of Union concern and make EU-funds available for this. This might be followed by a second plan of the same kind, etc. Only after the successful implementation of several such projects should there be new IAS of Union concern agreed.

6. Final remark

There is consensus that IAS of Union concern require action at Union level. However, the initiative of taking such action at Union level must come from the EU institutions [62]. The invitation to cooperate and concert actions between Member States (Article 22) was a flop. Member States continued their national, regional or local activities with regard to IAS as before. Initiatives by the EU will have to make EU funding available in order to bring an added value to the fight against IAS.

The main message is that in order to reach results, within the EU or at international level, close cooperation between neighbouring countries is necessary. It is not sufficient to leave the implementation and effective application of international agreements or of EU legislation to the goodwill of the countries concerned. The Conference of the Parties to the CBD as well as the European Commission will therefore have to do more to ensure an effective application of the existing provisions.

Author details

Ludwig Krämer
Independent Scientist

*Address all correspondence to: kramer.ludwig@hotmail.com

References

[1] See IUCN: "Invasive alien species are species that are introduced, accidentally or voluntarily, outside of their natural geographic range and become problematic", iucn.org/theme/species/our-work/invasive-species (accessed 29-9-2020).

[2] Convention on Biological Diversity, Article 8(h).

[3] CBD, Aichi Target no.9: "By 2020, invasive alien species and pathways are identified and prioritized, priority species are controlled or eradicated, and measures are in place to manage pathways to prevent their introduction and establishment".

[4] EU Commission, COM (2011) 244: "Our life insurance, our natural capital: an EU biodiversity strategy to 2020", Target 5: "By2020, invasive alien species and their pathways are identified and prioritized, priority species are controlled or eradicated, and pathway are managed, to prevent the introduction and establishment of new IAS".

[5] Regulation (EU) No 1143/2014 of the European Parliament and of the Council of 22 October 2014 on the prevention and management of the introduction and spread of invasive alien species, EU Official Journal (OJ) 2014, L 317, p.35.

[6] Decision 1600/2002 laying down the Sixth Community Environment Action Programme, OJ 2002, L 242 p.1, Article 6(2)(a): "(The Community will develop) measures aimed at the prevention and control of invasive alien species and alien genotypes".

[7] See in particular M. Kettunen - P. Genovesi - S. Gollasch- S.Pagad - U.Starfinger - P. Ten Brink - C. Shine: Technical support to EU strategy on invasive alien species (IAS) - assessment of the impacts of IAS in Europe and the EU. Institute for European Environmental Policy. Bruxelles2009. This study estimated the annual damage caused by IAS in the EU at about 12 billion euro.

[8] Commission Staff Working Document, Impact assessment, SWD (2013) 321.

[9] Commission, COM (2013) 620.

[10] European Parliament, Position on the Commission proposal for a regulation on IAS of 16 April 2014, Document TC1_COD(2013) 6307. The most important amendments suggested by the Parliament were the fixing of a deadline for the establishment of a list on IAS of EU concern; the inclusion of animal welfare considerations in the text; the deletion of limiting the future EU list of IAS to 50 species; and specific provisions for species that were IAS in some, but considered indigenous in other Member States.

[11] European Economic and Social Committee, Opinion of 11 June 2014, OJ 2014, C 177, p.84. The EU Committee of the Regions did not issue an Opinion.

[12] Commission Delegated Regulation 2018/698 with regard to risk assessments in relation to invasive alien species, OJ 2018,L 174, p.5; Commission Implementing Regulation 2016/145 adopting the format of the document serving as evidence for the permit issued by the competent authorities of the Member States allowing establishments to carry out certain activities pursuant to Regulation 1143/2014, OJ 2016, L 30,p.1; Commission Implementing Regulation 2017/1454 specifying the technical formats for reporting by the Member States pursuant to Regulation 1143/2014, OJ 2017, L 208, p.15. See generally on the implementing powers of the Commission under Regulation 1143/2014 (fn.5) Recital 31 of that Regulation.

[13] Commission Implementing Regulation 2016/1141, OJ 2016, L 189, p.4; the list contained 37 IAS species.

[14] Commission Implementing Regulation 2017/1263, OJ 2017, L 182, p.37; the list contained 12 IAS species. Commission Implementing Regulation 2019/1262, OJ 2019, L 199, p.1; the list contained 17 IAS species.

[15] Commission, SWD (2013) 321 (fn.8), p.18.

[16] Commission, SWD (2013) 321(fn.8), p.7.

[17] The six chapters are: I. General provisions (Articles 1 to 6); II. Prevention (Articles 7 to 13); III. Early detection and rapid eradication (Articles 14 to 18); IV. Management of IAS that are widely spread (Articles 19 to 20); V. Horizontal provisions (Articles 21 to 23); VI. Final provisions (Articles 24 to 33). 38 Recitals explain, justify and comment the different articles.

[18] Treaty on the Functioning of the European Union, Article 288.

[19] See on the situation prior to 2014 for example J. Vandekerkhove - A.C. Cardoso: Online information systems with alien species occurrence records in Europe; coverage, complementarity and compatibilitiy. European Commission, Joint Research Centre. Luxemburg 2011.

[20] See the definitions in Article 3 of Regulation 1143/2014 (fn.5).

[21] Regulation 1143/2014 (fn.5), Article 22 mentions situations, where Member States share the same maritime sub-region, the same biogeographical region, share borders or river basins or have other "common concerns".

[22] This provision does not exclude, though, Commission's initiatives to ensure cooperation without a formal invitation.

[23] Regulation 1143/2014 (fn.5), Article 10.

[24] *Ibidem*, Articles 11 and 12.

[25] *Ibidem*, Article 12(3).

[26] *Ibidem*, Article19(5).

[27] *Ibidem*, Article27.

[28] *Ibidem*, Article28.

[29] See for the last two groups Commission, ec.europa.eu/environment/nature/invasivealien/index_en.htm (consulted 5-6-2020).

[30] Under the Regulation, Member States shall report to the Commission on the taking of emergency measures (Article10), the establishment of a list of IAS of Member State concern (Article 11), their action plans for pathways (Article13(5), the early detection of IAS (Article 16), the measures taken to combat early detected IAS (Article 17(1) and(4)), decisions not to apply rapid eradication measures (Article 18),the adoption of more stringent measures (Article 23), a detailed implementation report (Article 24) and their system of sanctions (Article 30).

[31] The EU LIFE programme (Regulation 1293/2013 on the establishment of a Programme for the Environment and Climate Action(LIFE), OJ 2013, L 347, p.185) financed, between 1992 and 2018, 123 projects on IAS, see ec.europa.eu/environment/life/project/Project/index.cfm?fuseaction=home.search&fid=16274911&cftoken=79af2267Sc8e9601-2E4B7050-B0B1-8670-42776375BD162288 (consulted 5-6-2020). 33 of these projects were financed in Italy, 28 in Spain, 12 in the United Kingdom, 11 in Portugal and 9 in France.

[32] According to Article 4(3)(d) of Regulation 1143/2014 (fn.5), a species

must be the subject of a risk assessment, before it can be inserted into the list of IAS of Union concern. These risk assessments must be realised by the Commission, eventually also by a Member State (Article 4(4)b)). For details of the risk assessment see Articles 5 of Regulation 1143/2014 and Regulation 2018/698 (fn.12). Particular importance is attached to the necessity of the risk assessment's compliance with the rules of the World Trade Organisation (Recitals 11 and 13 to Regulation 1143/2014).

[33] See K. Tsiamis a.o.: Baseline distribution of invasive alien species of Union concern. European Commission, Joint Research Centre, Luxemburg 2017 (Tsiamis 2017); K.Tsiamis - E.Gervasini a.o.: Baseline distribution of species listed in the 1st update of IAS of Union concern. Luxemburg 2019 . These publications cover 49 of the 66 IAS of Union concern.

[34] See in particular the European Alien Species Information Network (EASIN), launched by the EU Commission, Joint Research System; S. Katsanevakis a.o: EASIN: supporting European policy and scientific research. Management of Biological Invasions 2015, p.147.

[35] See Article 193 of the Treaty on the Functioning of the European Union and Regulation 1143/2014 (fn.1), Article 23.

[36] For their outermost regions, though, Member States were obliged to establish IAS lists of concern for those regions, Article 6 of Regulation 1143/2014. These outermost regions, which are part of the EU, are Martinique, Mayotte, Guadeloupe, French Guiana, Réunion (France), Azores, Madeira (Portugal) and the Canary Islands (Spain).

[37] Regulation 1143/2014 (fn.5), Articles 4(1) and 27.

[38] *Ibidem*, Article 4(6).

[39] See for details Regulation 1143/2014 (fn.5) Article 27 and Regulation 182/2011,OJ 2011, L 55, p.13.

[40] See for details Regulation 1143/2014 (fn.5), Article 7.

[41] *Ibidem*, Articles 8 and 9. Transitional provisions were foreseen for non-commercial owners of IAS (Article 31) and on the liquidation of commercial stocks of IAS (Article 32)

[42] *Ibidem*, Article 10.

[43] On request of the European Ombudsman, scientific experts have now to be registered in the EU transparency register. However, this does not yet help with regard to the composition of the scientific expert group. Even data protection reasons do not justify that the names of the experts are not made public.

[44] Commission, Environmental Implementation Review 2019, COM (2019) 149, p.6. The Member States in question were Austria, Belgium, Cyprus, Greece, Ireland, Portugal, Romania, Slovakia, Sweden and the United Kingdom.

[45] *Ibidem*. The Member States concerned were Czechia, Denmark, Germany, Greece, France, Hungary, Portugal and Spain.

[46] According to Article 24, Member States have to report on their surveillance system (Article 14), their control system (Article 15), the distribution of IAS in their territory, IAS of Member State concern, their action plans (Article 13), management measures (Articles 17 and 19) and their effectiveness, permits that were issued, information of the public, and, "if available", cost of the measures.

[47] Commission, rod.eionet.europa.eu/obligations/727/deliveries (consulted

9-6-2020). The report from Latvia was not accessible, though. The reports follow the orientation of Commission Regulation 2017/1454 (fn.8). As the United Kingdom left the EU, data on that country are not included in the following lines.

[48] Commission, easin.jrc.ec.europa.eu/notsys/PUB/search/GetResults (consulted 9-6-2020).

[49] Regulation 2017/1454 (fn.12).

[50] Commission, SWD (2019) 112 to 139, doubted, whether the reports from Bulgaria, Hungary, Romania and Denmark were not influenced by poor data availability. To these, certainly the Lithuanian report has to be added, which indicates with regard to a ten IAS of Union concern that it was "unknown", whether these species existed in the national territory. The report for Greece even indicated that with regard to 43 IAS it is "currently unknown", whether the IAS exists in the national territory. See for both countries rod.eionet.europa.ec/obligations/727/deliveries (Lithuania, Greece) (consulted 10-6-2020).

[51] See C. Carboneras a.o.: A prioritised list of invasive alien species to assist the effective implementation of EU legislation. Journal of Applied Ecology 2018, p.539.

[52] Commission, notsys (fn.48).

[53] *Ibidem*. In detail: Germany 29 notifications, Netherlands 12, Ireland 8, Luxemburg 6, Spain 5, Belgium 5, Denmark 3, Slovenia 3, Croatia 3, Sweden 2, Italy 2, Austria 1, Czechia 1, Portugal 1, Hungary 1 and France 1. Some notifications concerned several populations. However, some Member State apparently adopted eradication measures on the basis of Article 17, without notifying the Commission, see for example the annual reports from Denmark (9 eradication measures) and France (4 eradication measures), Commission, rod.einet.europa.ec/obligations/727/deliveries (Denmark, France) (consulted 10-6-2020).

[54] The notifications concerned: Vespa velutina nigrithorax (14 notifications); *Oxyura jamaicensis* (10); *Muntiacus reevesi* (8); *Myocastor coypus* (6); *Threskiornis aethiopicus* (5); Cabomba carolina (4); *Procyon lotor* (4); *Nasua nasua* (3); *Ludwigia peploides* (3); *Procambarus fallax* (3); *Pueraria montana* (2); *Eichhornia crassipes* (2); *Lagarosiphon major* (2); *Persicaria perfoliata* (2); *Procambarus clarkii* (2); *Tamias sibiricus* (1); *Corvus splendens* (1); *Sciurus carolinensis* (1); Hydrocotyle ranuncaloides(1); *Lysichiton americanus* (1);Lithobathes catesbeianus (1); Alopochen aegyptianus (1); Gymnocoronis spilanthoides (1); Myriophyllum heterophyllum(1).

[55] Greece reported that 6 of the 49 IAS were present in its territory, Bulgaria 12, Romania 13 and Cyprus 3 IAS (see the reports fn.48).

[56] No presence was signalled of *Corvus splendens*, Microstegium viminieum, *Nyctereutes procyonoides*, Parthernium hystrophorous, *Persicaria perfoliata* and *Sciurus niger*. Where a Member State reported that it did not know, whether a species was present in its territory, the species was counted as not present.

[57] National lists of IAS exist in Croatia, Denmark, Estonia, France, Ireland, Lithuania, Poland, Slovakia and Spain, see Commission (fn.48), national reports.

[58] See fn.55. The presence of the Asian hornet was signaled by four Member States, that of the Ruddy duck by 12 and of Coypus by 15 Member States.

[59] See Tsiamis a.o. 2017, (fn.33), p.31 and p.49.

[60] The countries are Belgium, France, Germany, Hungary, Italy, Netherlands and Spain. Furthermore, 13 countries adopted measures with regard to less than half of the IAS of Union concern present in their territory, and 28 of the 43 IAS of Union concern, on which Member States reported, were subject to measures in less than half of the Member States; see table above.

[61] This might be different for aquatic animal IAS, which, until now, were not particularly addressed by the EU. Indeed, the authorities dealing with nature conservation (including fighting IAS) and water protection are separated in most Member States. Aquatic IAS of Union concern would therefore have to be managed by different administrations.

[62] Such Union action will have to include non-EU countries such as Switzerland, Norway, Liechtenstein and the United Kingdom.